Energy Conservation Guidebook

Third Edition

by
Dale R. Patrick
Stephen W. Fardo
Ray E. Richardson
Brian W. Fardo

River Publishers

Routledge
Taylor & Francis Group

LONDON AND NEW YORK

Published 2020 by River Publishers

River Publishers

Alsbjergvej 10, 9260 Gistrup, Denmark

www.riverpublishers.com

Distributed exclusively by Routledge

4 Park Square, Milton Park, Abingdon, Oxon OX14 4RN

605 Third Avenue, New York, NY 10017, USA

First issued in paperback 2023

Library of Congress Cataloging-in-Publication Data

Patrick, Dale R.

Energy conservation guidebook / by Dale R. Patrick, Stephen W. Fardo, Ray E. Richardson, Brian W. Fardo. -- Third edition.

pages cm

Origninal 1993 edition by Steven R. Patrick, Dale R. Patrick, Stephen W. Fardo.
Includes index.

ISBN 0-88173-716-X (alk. paper) -- ISBN 978-8-7702-2319-5 (electronic) -- ISBN 978-1-4822-5569-0 (Taylor & Francis distribution: alk. paper) 1. Energy conservation--Handbooks, manuals, etc. I. Fardo, Stephen W. II. Richardson, Ray E., 1961- III. Fardo, Brian W. IV. Title

TJ163.3.P38 2014
696--dc23

2014011314

Energy conservation guidebook, 3rd edition / by Dale R. Patrick
First published by Fairmont Press in 2014.

Routledge is an imprint of the Taylor & Francis Group, an informa business

Publisher's Note
The publisher has gone to great lengths to ensure the quality of this reprint but points out that some imperfections in the original copies may be apparent.

ISBN-978-1-4822-5569-0 (pbk)
ISBN-978-1-4822-5569-0 (hbk)
ISBN-978-8-7702-2319-5 (online)
ISBN-978-1-0031-5188-3 (ebook master)

While every effort is made to provide dependable information, the publisher, authors, and editors cannot be held responsible for any errors or omissions.

The views expressed herein do not necessarily reflect those of the publisher.

Table of Contents

Chapter 5

Summer Air-Conditioning Systems/
Saving Natural Resources 135

Chapter 6

Lighting Systems/Improved Efficiency 199

Chapter 7

Chapter 8

Chapter 9

Preface

Efficient energy management and effective conservation procedures have been very important considerations for our society for many years. An oil embargo in the 1970s and early 1980s brought about a new awareness of energy conservation. Because of various factors like loss of tax credits and efficiency standards imposed by the government, public interest dropped considerably in regard to energy conservation. A revival in energy conservation among the general public occurred following the Persian Gulf War in the early 1990s. What does the 21st century hold? Conflicts in the Middle East, high prices for petroleum, and increasing population worldwide will all be significant influences on energy and its' conservation.

Energy Management and Conservation, 3rd Edition, provides a very practical discussion of how energy can be managed and saved in most types of buildings. This edition not only updates the previous edition, but adds an updated chapter concerning energy cost reduction/going green to improve the environment.

The authors of this book have written several books that use the systems approach. This is a method that helps the reader to understand how related subjects "fit together" in a common format. Through the use of the systems approach, the reader will be able to grasp how different parts of a building fit together to form a unit that uses energy efficiently.

This book should be of interest to a wide variety of individuals. Some of these include vocational-technical schools, teachers, industrial training managers, building maintenance personnel, business owners, and homeowners.

Energy Management and Conservation, 3rd Edition, provides a thorough and practical discussion of the operation of systems that are found in most types of buildings. Each system is discussed with energy management and conservation in mind—going green to save money and improve the environment. There are many ways to manage a building to accomplish efficient energy conservation. Many of the chapters have checklists at the end to summarize ways of conserving energy which relate to that chapter.

In the text, discussion is centered around the efficiency of a particular system. Procedures to modify and maintain existing equipment or systems are given to the reader. The chapters of this book develop a basic, easy-to-understand explanation of the operating systems of a building. The chapters are organized in the following order:

1.   Introduction (brief system overview).
2.   Descriptive content (main text of chapter).
3.   Energy conservation checklist (condensed for easy reference).

The authors would like to thank all those who helped in the preparation of this manuscript. Many companies supplied technical data, illustrations, and photographs. Their cooperation is greatly appreciated.

In addition, we would like to thank Mr. Mark Barron, Eastern Kentucky University Building Control Systems Technician, for assisting with updated photos and technical information to support this revision.

Dale R. Patrick
Stephen W. Fardo
Ray E. Richardson
Brian W. Fardo

Chapter 1

Introduction/
Green Energy Concept

Energy management and conservation are the keys to using fuel and electrical energy in the most efficient way. Proper energy management can lead to big savings on the operating costs of a building. If fuel and electrical energy consumption are reduced, money will be saved as a result. Many residential, industrial, and commercial buildings have already undergone changes that have resulted in the savings of both energy and money. Any building can be made more energy efficient when proper energy management procedures are applied.

REASONS FOR ENERGY MANAGEMENT

Good energy management in buildings will also help to conserve our valuable natural resources. This is often referred to as "going green." (See Chapter 13.) Money savings and conservation are the two major benefits of energy management. A few other important results are less dependence on imported oil and other sources plus the longer life of some equipment. This book deals primarily with energy management and conservation in existing buildings. However, there are many suggestions in the book that should be considered in both the design of new buildings and the remodeling of existing buildings.

There are many inexpensive changes that can be made in existing buildings which will save energy and money and also improve the environment. Many or our existing buildings were constructed prior to the early 1970s, when energy conservation was not a national problem or a major financial consideration. Buildings constructed prior to the energy crisis of the early 1970s and early 1980s were built with a certain amount of energy efficiency in mind. After oil prices calmed down in the latter 1980s, energy conservation was again paramount only to certain people.

Following the Desert Storm conflict in the 1990s, people were expecting oil prices again to drop, but just the opposite happened. Climbing prices landed this country in a recession and once again energy conservation became a priority with some individuals. This little history lesson has shown that the public normally spends both money and time educating themselves about energy only during times of crisis. One good thing that has come out of these different crises is that most of the public is more energy conscious because they at least know energy savings might render them monetary savings. The global impact has been to improve the environment.

OVERVIEW OF ENERGY MANAGEMENT

Energy Conservation Guidebook, 3rd Edition, is organized into chapters which discuss the "systems" or parts of typical buildings. These systems include the building structure (Chapter 3). The structure of a building, such as the one shown in Figure 1-1, is sometimes called the *building envelope.* The building shown has solar window film for energy conservation. Simple modifications of the building structure, such as adding insulation, can provide energy savings with a small financial investment which will pay for itself in a short period of time. Considerations for determining the payback period of any equipment or material purchased are also discussed in this book.

Figure 1-2 shows some heating systems that are in common use in buildings. An overview of the many types of heating systems in use today is presented in Chapter 4. In most geographic areas, heating systems consume a greater amount of energy than any other building system. The cooling system of a building is often integrated with the heating system. Cooling systems or air-conditioning and ventilating systems are discussed in Chapter 5. The combined heating and cooling system is usually referred to as the HVAC (heating, ventilating, and

Figure 1-1. The building structure. (Courtesy of Madico Co.)

air-conditioning) systems of a building. In the past, HVAC systems were generally designed with initial cost as the primary consideration. It is now important to consider the energy efficiency of the system to reduce long-term energy costs.

Lighting systems, such as those shown in Figure 1-3, are another important part of buildings. The electrical energy used to power lights can easily be reduced. Lighting systems are used not only to provide sufficient light inside buildings but also for beauty and security on the outside of buildings. Simple modifications of lighting systems can greatly reduce the energy used while still providing quality and illumination needed for various purposes. There have been several recent developments and research findings in the lighting industry which can provide reduced energy consumption.

(a)

(b)

Figure 1-2. Heating systems: (a) boiler for a hot-water heating system; (b) electric steam boiler; (c) electric furnace. (*a, courtesy of A.O. Smith Co.; b, courtesy of Patterson-Kelley Co., Div. of Harsco Corp.; c, courtesy of Lennox Industries, Inc.*)

(c)

Figure 1-3. Indoor lighting systems. (*Courtesy of Armstrong Cork Co.*)

Figure 1-4 shows a water system that is part of a building. Domestic hot-water systems consume a significant amount of energy in several types of buildings. Cold water from drinking fountains is also a consideration in total energy usage. The condition of the water system of a building has an important impact on energy-conscious operation of the system.

Figure 1-4. Water system; industrial-commercial water heater. (*Courtesy of Patterson-Kelley Co.***)**

The electrical power system of a building is summarized in Chapter 8. Many parts of a building use electrical power, so it is important to assure that all systems operate efficiently. Electrical power bills, such as the one illustrated in Figure 1-5, have steadily increased. People are becoming more conscious of energy savings as a result of increased costs. Proper electrical design and electrical use management in a building can provide long-term financial savings. The energy cost calculator shown in Figure 1-5 can be used for estimating electrical energy cost.

Chapter 9 deals with solar energy systems. Although solar energy systems have been much publicized, their immediate applications for energy conservation in existing buildings are limited. However, such applications are becoming more realistic as energy costs increase. Active and passive solar system design of buildings can aid in heating, cooling, and domestic hot-water systems. Figure 1-6 shows an illustration of a solar energy monitoring system. These instruments can be used to monitor solar installations and evaluate sites for potential solar applications. Most solar energy systems are used to supplement existing heating and cooling systems rather than to supply 100% of a building's energy needs.

Energy control and measurement systems are discussed in Chapter 10. Various types of control and measurements are accomplished in most buildings. A typical type of measuring device is the kilowatthour meter used to measure electrical energy usage. Some electrical energy monitoring equipment is shown in Figure 1-7. Energy control and measurement play a significant role in accomplishing energy conservation.

Chapter 11 is a capstone for the preceding chapters. Energy management systems, such as the computerized unit shown in Figure 1-8a, were used in the 1990s. Modern systems are now being used to control energy-consuming equipment in large buildings. The screen shot shown in Figure 1-8b can run anywhere, controlled by a networked computer or tablet computing device. The primary emphasis of Chapter 11 is to show how an energy management program for a business can be developed. This chapter also stresses the importance of energy conservation with techniques of calculating actual financial costs and savings. Energy conservation can have significant economic implications for businesses.

A building can be inspected very easily to see what can be done to save energy. A method of checking a building has become known as an *energy audit*. An energy audit can be done by any person who is familiar with a building. It can also be done to a higher degree of sophistication by trained professional people. This book can be used as a reference for performing an energy audit for a building. The method used to make an energy audit is discussed in Chapter 11.

Chapter 12 introduces alternative forms of energy—geothermal, wind, tidal, biomass, magnetohydrodynamics, and nuclear power. This discussion stimulates thought about potential alternative systems. Solar power is an alternative energy source, but a viable, and is discussed in detail in Chapter 9. Each system discussed in this chapter has many potential problems. No matter how serious, experimentation must be conducted to assure that electrical power can be produced economically. Our technology depends on low-cost electrical power.

Figure 1-5. Electrical power bill and cost calculator. *(Courtesy of Phillips Lighting Co.)*

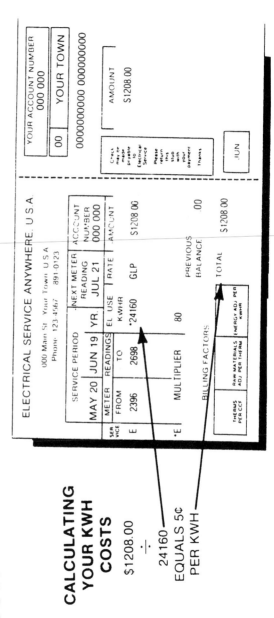

ENERGY COST CALCULATOR GUIDE...

ONE WATT • ONE YEAR

COST PER KILOWATT HOUR IN CENTS

HOURS USED PER DAY	2	2.5	3	3.5	4	4.5	5	5.5	6	6.5	7	7.5	8	8.5	9	9.5	10	10.5	11
8	5	6	7	9	10	11	12	14	15	16	17	19	20	21	22	24	25	26	27
10	6	8	9	11	12	14	16	17	19	20	22	23	25	27	28	30	31	33	34
12	7	9	11	13	15	17	19	21	22	24	26	28	30	32	34	36	37	39	41
14	9	11	13	15	17	20	22	24	26	28	31	33	35	37	39	41	44	46	48
18	11	14	17	20	22	25	28	31	34	37	39	42	45	48	51	53	56	59	62

ANNUAL COST PER CONSUMED WATT BASED ON SIX (6) DAY USAGE*   *Add 1/6 for 7 days   Subtract 1/6 for 5 days

EXAMPLE: 5¢ per KW x 12 hours of daily usage = 19¢ per watt, per year

COST PER WATT   24 HOURS • 7 DAYS • 52 WEEKS (8760 HOURS)

18	22	26	30	35	40	43	48	53	57	61	65	70	75	78	83	88	92	96

One (1) Kilowatt Hour (KWH) = 1,000 Watts used one hour

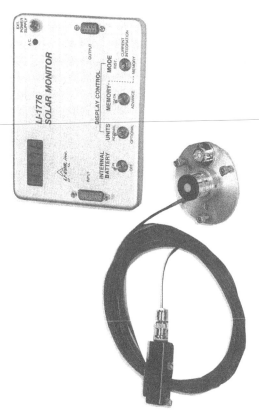

Figure 1-6. Solar energy monitoring system. (Courtesy of Li-Cor, Inc.)

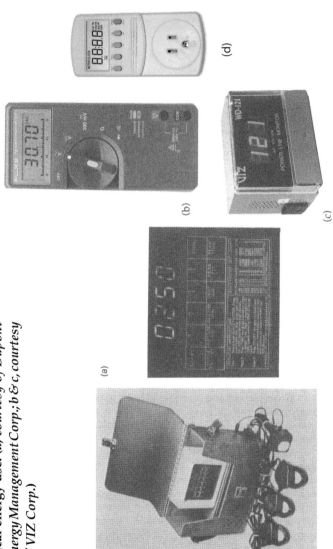

Figure 1-7. Meters used to monitor electrical energy use. *(a, courtesy of Dupont Energy Management Corp.; b & c, courtesy of VIZ Corp.)*

Figure 1-8a. Energy management system. (*Courtesy of Honeywell, Inc.*)

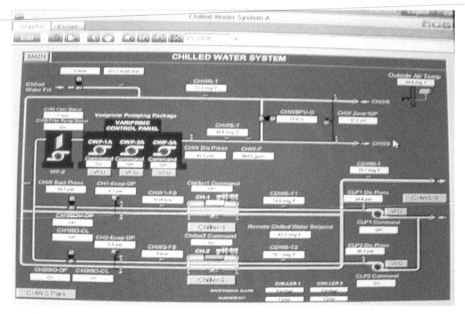

Figure 1-8b. Energy management system. (*Courtesy of Honeywell, Inc.*)

Chapter 13, a discussion of techniques for energy cost reduction/ going green, offers methodologies for reducing energy costs from the uncomplicated, inexpensive to system-wide changes. These changes also impact environmental quality.

THE SYSTEMS CONCEPT

Much reference is made in this book to the *systems* concept. This concept allows us to discuss some rather complex systems in a simplified manner by looking at an entire operational system or unit rather than only its parts. This method is used to present the chapters of the book and make them easier to understand.

For many years, people have worked with jigsaw puzzles as a source of recreation. This type of puzzle contains a number of discrete parts that must be properly placed together to produce a picture. Each part then plays a specific role in the finished product. When a puzzle is first started, it is difficult to imagine the finished product unless one sees a representative picture.

When one studies a complex field such as energy conservation by using discrete parts, it poses a problem that is somewhat similar to a jigsaw puzzle. It is difficult to determine the role that each part plays in the operation of a complex system.

The systems concept will serve as our "big picture" in the study of energy conservation. In this approach, we will initially divide a system into a number of parts. The role played by each part will then become more meaningful in the operation of the overall system. After the function of each part has been established, discrete component operation related to each block will then become more relevant. Through this approach one should soon be able to see how the "pieces" of the energy conservation field fit together in a more meaningful order.

System Functions

The word *system* is commonly defined as "an organization of parts that are connected together to form a complete unit." There are a wide variety of different systems used today. An electrical power system, for example, is needed to produce electrical energy and distribute it to each part of a building. Each system obviously has a number of unique features or characteristics that distinguish it from other systems. More important, however, there is a common set of parts found in most systems. These parts play the same basic role in all systems. The terms *energy source, transmission path, control, load,* and *indicator* are used in this book to describe the various system parts. A block diagram of the parts of the system is shown in Figure 1-9.

Each block of a basic system has a specific role to play in the overall operation of the system. This role becomes extremely important when a detailed analysis of the system is to take place. Hundreds and even thousands of discrete components are sometimes needed to achieve a specific block function. Regardless of the complexity of the system each block must still achieve its function. Regardless of the complexity, for the system to be operational each block must still achieve its function. Being familiar with these functions and being able to locate them within a complete system is a big step in understanding the operation of the entire process.

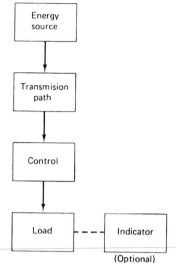

Figure 1-9.
The systems concept.

The energy source of a system is responsible for converting energy of one form into something useful. Heat, light, sound, chemical, nuclear, and mechanical energy are considered as primary sources of energy. A primary energy source usually goes through an energy transformation before it can be used in an operating system.

The transmission path of a system is somewhat simplified when compared with other system functions. This part of the system simply provides a path for the transfer of energy. It starts with the energy source and continues through the system to the load. In some cases, this path may be a feed line, electrical conductor, light beam, or pipe connected between the source and the load plus a return line from the load to the source. There may also be a number of alternative or auxiliary paths within the complete system.

The control section of a system is by far the most complex part of the entire system. In its simplest form, control is achieved when a system is turned on or off. Control of this type can take place anywhere between the source and the load device. The term full control is commonly used to describe this operation. In addition to this type of control, a system may also employ some type of partial control. Partial control usually causes some type of an operational change in the system other than an on or off condition. Changes in electrical current, pressure, light intensity, and airflow are some of the system alterations achieved by partial control.

The *load* of a system refers to a specific part or number of parts designed to produce some form of work. The term *work*, in this case, occurs when energy goes through a transformation or change. Heat, light, chemical action, sound, and mechanical motion are some of the common forms of work produced by a load device. As a general rule, a very large portion of all energy produced by the source is converted by the load device during operation. The load is typically the most prevalent part of the entire system because of its obvious work function.

The *indicator* of a system is designed primarily to display certain operating conditions at various points throughout the system. In some systems the indicator is an optional part, whereas in others it is an essential part in the operation of the system. In the latter case, system operation and adjustments are usually critical and are dependent upon specific indicator readings. The term *operational indicator is* commonly used to describe this application. Test indicators are also needed to determine different operating values. To make measurements the indicator is attached to the system only temporarily. Test lights, panel meters, oscilloscopes, chart recorders, digital display instruments, and pressure gauges are some of the common indicators used in this capacity.

Building Operating Systems

The number of different systems used in buildings today is quite large when we consider the wide variety of different functions that are being accomplished. Each building has a unique application. Many energy sources, such as heat, light, electrical, and mechanical energy are needed for a building. The types of buildings in existence today include residential, commercial, and industrial. There are many different classifications of each type of building.

Electrical System Examples

Nearly all of us have had an opportunity sometime to use a flashlight. This device is designed to serve as a light source in an emergency or to provide light to unusual places. In a strict sense, flashlights can be classified as portable electrical systems. They contain the four essential parts needed to make a system. Figure 1-10 is a cutaway drawing of a

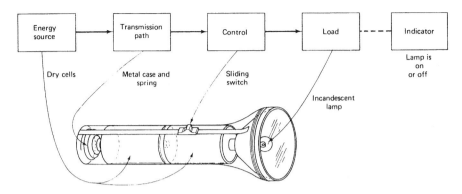

Figure 1-10. Cutaway drawing of a flashlight.

flashlight with each component part shown.

The battery of a flashlight serves as the energy source of the system. Chemical energy of the battery must be changed into electrical energy before the system becomes operational. The energy source of a flashlight is an expendable item. It must be replaced periodically when it loses its ability to produce electrical energy.

The transmission path of a flashlight is commonly achieved by a metal case or through an electrical conductor strip. Copper, brass, and plated steel are frequently used to achieve this function.

The control of electrical energy in a flashlight is achieved by a slide switch or pushbutton switch. This type of control simply closes or opens the transmission path between the source and the load device. Flashlights are designed to have full control capabilities. This type of control is achieved manually by the person operating the system.

The load of a flashlight is a small incandescent lamp. When electrical energy from the source is forced to pass through the filament of the lamp, the lamp will produce a bright glow. Electrical energy is then changed into light energy. A certain amount of work is achieved by the lamp when this energy change takes place.

Flashlights do not ordinarily use a specific indicator as part of a system. Operation is indicated, however, when the lamp produces light. In a strict sense, we could say that the load of this system also acts as an indicator. In some electrical systems, the indicator is an optional system part.

Another example of a system is the electrical power system that

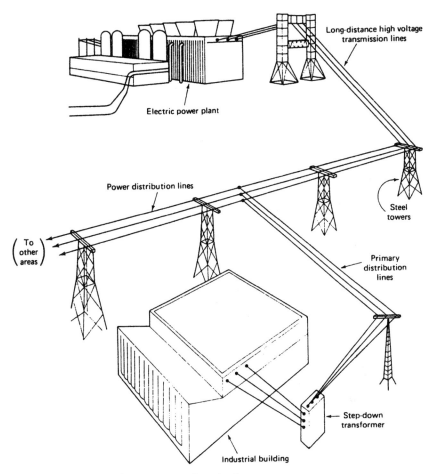

Long-distance high voltage transmission lines

Electric power plant

Power distribution lines

To other areas

Steel towers

Primary distribution lines

Step-down transformer

Industrial building

Figure 1-11. Electrical power system.

supplies energy to residential, commercial, or industrial buildings. Figure 1-11 shows a sketch of a simple electrical power system. These systems are discussed in detail in Chapter 8.

The energy source of an electrical power system is much more complex than that of the flashlight that was discussed earlier. The source of energy may be derived from coal, oil, natural gas, atomic fuel, or moving water. This type of energy is needed to produce mechanical energy, which in turn develops the rotary motion of a turbine. Massive alternators are then rotated by the turbine to produce alternating-current (AC) electricity. The energy-conversion process of this particular system is quite involved from start to finish. Its function is the same, however, regardless of its complexity.

In an electrical power system, the transmission path is achieved by a large number of electrical conductors. Copper wire and aluminum wire are used more frequently today than any other type of conductor. Metal, water, the earth, and the human body can all be made to conduct energy when contact is made with certain parts of an electrical power system. To avoid an electrical shock, extreme caution must be observed when working with an operating electrical power system. Ordinarily, interior electrical conductors are insulated to prevent shock hazards.

The transmission path of an electrical power system often becomes very complex. They are usually referred to as electrical power distribution systems.

The control function of an electrical power system is achieved in a variety of different ways. Full control, for example, is accomplished by three types of circuit-interrupting equipment. These include switches, circuit breakers, and fuses. Each piece of equipment must be designed to pass and interrupt specific values of current. Partial control of an electrical power system is achieved by various types of circuits. To minimize power losses in an electrical system transformers are used at strategic locations throughout the system. These partial control devices are designed initially to step up the source voltage to a higher value. Through this process, the source current is reduced in value proportionally. Since the power loss of a transmission line is based on the amount of current, power losses can be reduced to a reasonable value through this method.

Transformers are also used to lower system voltages to usable values near the load. This action of a transformer is described as its step-down function. When the source voltage is reduced to a lower value the current is increased in value proportionally. Through the use of transformers, transmission-line losses can be held to a minimum, thus causing increased system efficiency.

The load of an electrical power system is usually quite complex. As a composite, it includes everything that uses electrical energy from the source. Ordinarily, the load is divided into four distinct types: residential, commercial, industrial, and other uses, such as street lighting. The composite load of an electrical power system is subjected to change hourly, daily, and seasonally.

The average person is probably more familiar with the load part of the electrical power system than with any of its other parts. This represents the part of the system that actually does work. Motors, lamps, electric ovens, welders, and power tools are some of the common load devices used. Loads are frequently classified according to the type of work they produce: light, heat, electromechanical changes, and chemical action.

The indicator of an electrical power system is designed to show the presence of electrical energy at various placed or to measure different electrical quantities. Panel-mounted meters, oscilloscopes, chart recording instruments, and digital display devices are some of the indicators used in this type of system today. Indicators of this type are designed to provide an abundance of system operating information.

System Summary

The systems concept is an orderly method that can be used to study the field of energy conservation in buildings. This idea describes a common organizational plan that applies to most systems. Each part of a system plays a similar role in all systems. The energy source, transmission path, control, load, and indicator are basic to all systems. An understanding of the basic system plan helps to overcome some of the complexities of different types of systems.

The energy source of a system is responsible for producing energy to be used by the system. Heat, light, sound, chemical, and mechanical energy are primary sources of energy.

The transmission path of a system provides a means by which energy can be passed from the source to other system parts. Light beams, electrical conductors, and pipes are typical transmission paths.

System control can be either full or partial. Full control is an on/off operation, whereas partial control adjusts or varies system values. Each system type usually has a number of unique control features. Temperature, light, mechanical motion, time, sound, electric current, hydraulic fluid flow, and air are controlled in various types of systems.

The load of a basic system is responsible for changing system energy into some other form of energy. Work occurs in the load when it causes a change or transformation of energy.

The indicator part of a system is designed to display or show certain operating conditions. Test indicators are temporarily attached to the system to locate faulty components. Operational indicators, by comparison, are permanently attached to a system to display critical operating values. Indicators are usually classified as optional system parts.

Chapter 2

Energy Basics/
Foundation for Understanding

INTRODUCTION

The advancement of science and technology has brought about a large number of very important changes in the basic structure of a building and the equipment that is used to keep it operational. Most building equipment has become somewhat complex and requires skilled personnel to keep it in operation. Technicians are called upon to analyze this equipment, maintain it in good operating order, and recommend energy conservation measures. A wide range of experience is needed in different areas to cope with these situations.

At one time, most building equipment could be placed into operation with a few simple tools and some good common sense. Today, however, a great deal of our building equipment involves some form of control that performs precise operations automatically. Building personnel must now be concerned with such things as evaluation procedures, calibration, instrumentation, and troubleshooting techniques to maintain this kind of equipment. In addition to this, there is an increased concern for such things as operational efficiency, preventative maintenance, and energy management.

Building equipment operation today is based on a number of very important fundamental principles. A person working with this equipment must have some understanding of these principles in order to work effectively. A great deal of this basic material will be a review of scientific principles for those readers who have studied the subject. In addition to this, there are some operating practices and technical principles that must be understood. As a general rule, these principles have been simplified by relating them to practical building applications. Areas of concern relate to molecular theory, heat, pressure, humidity, work, power, and energy.

21

MATTER

In the world about us, we find a wide variety of things, such as air, water, wood, metal, stone, paper, and living things. These substances are all common examples of matter. Although matter exists in many different forms, it has two very basic properties with respect to its weight and the space that it occupies.

The quantity of matter that a body contains is called its *mass*. Since there is twice as much liquid in a gallon as there is in a half-gallon, the gallon has twice the mass of the half-gallon.

All mass in a sense is pulled toward the center of the earth by the force of gravity. This downward pull exerted by gravity determines the weight of a body. *Weight* is directly proportional to its mass and inversely proportional to the square of its distance from the center of the earth. A body with a great deal of mass has more weight than one with less mass. When a given body moves farther away from the center of the earth, its weight decreases. For this reason a certain item will weigh less on a high mountain than it would on the coast at sea level.

States of Matter

All matter, regardless of where it exists in the universe may appear in any one of these distinct forms or states: solids, liquids, or gases. Each state has its own unique characteristic that distinguishes it from the others. In its *solid* state, matter has a definite volume and physical shape. Representative solids are stone, glass, metal, wood, and paper. *Liquids* are quite different to the extent that they have a definite volume but do not have a specific shape. They conform to the shape of the container in which they are placed. Water, oil, alcohol, and gasoline are common examples of liquids. *Gas* differs from the others by not having a definite volume or shape. Typical examples of gas are air, oxygen, hydrogen, and neon.

Many substances may exist in all three states of matter, depending upon the temperature. If the temperature of water is below 32°F (0°C), it will appear in a solid state as ice. At room temperature, water is normally in a liquid state. Increasing the temperature of water to 212° F (100° C) causes it to change into steam or to the gaseous state.

Composition of Matter

Scientists generally believe that all matter is composed of tiny particles called *molecules*. All molecules of a particular substance are assumed to be alike, whereas those of another substance take on a different form. Molecules are so small in size that it takes 1000 or more of them sitting side by side to be visible on our best microscopes. It has been estimated that a 1-quart container of any gas under ordinary conditions of temperature and pressure contains approximately 25×10^{21}, or 25,000,000,000,000,000,000,000, molecules.

A molecule is defined as the smallest particle into which matter can be divided and still retain its original chemical identity. Thus, a molecule of water is considered to be the smallest quantity of water that can exist and still be classified as water. Scientists now believe that molecules themselves are composed of smaller particles known as *atoms*. An atom, by itself, is an independent particle that does not possess properties of the original material from which it was obtained. There are 92 different kinds of atoms found naturally, with several more being produced by nuclear bombardment. A molecule of water, which is classified chemically as H_2O, is composed of two parts hydrogen, H_2, and one part oxygen, O. The physical state of hydrogen and oxygen by themselves do not have the same properties as water.

It is important to note that atoms are also composed of smaller or subatomic particles. These are called electrons, protons, and neutrons. An *electron* holds a negative electrical charge, whereas a *proton* possesses a positive charge. *Neutrons* are electrically neutral and have no charge. Electricity is based upon the flow or movement of electrons within electrical conductors.

Molecular Motion

We know that matter exists throughout the universe in three distinct states. Water, for example, may be liquid, frozen into a solid state, or vaporized into the gaseous state. In each form, it is still composed of identical molecules. The primary difference in the physical states of matter can be explained by the relative position of molecules in a material and the freedom of their motion. We know today that these molecules are continuously moving about in all directions and at different velocities.

The molecules of matter when it is in a solid state are believed to vibrate or oscillate around fixed positions or locations. Individual molecules have a unique tendency to cling together by a force called *cohesion.* This force does not permit molecules to move very far away from their original position. As a result of this condition solids have a tendency to take on a definite shape.

In liquids, molecules are not held together firmly in a rigid pattern. The cohesion force between individual molecules, however, still exists. The resulting space between molecules is somewhat greater than that of a solid material. This unique difference in a liquid causes individual molecules to have more freedom in their movement. They have a greater tendency to slip over each other and to move around with case. As a result of this condition, molecules in a liquid do not remain in fixed positions, which causes the material to be in a constant state of change. Liquids do not have a special shape but, rather, conform to the dimensions of the container into which they are placed.

In gases, individual molecules are spread apart a great deal more than their liquid- and solid-state counterparts. This is exemplified by the fact that 1 cubic foot of water will expand into 1600 cubic feet of steam when it changes state. Steam molecules continue to be of the same consistency, the only difference being the space between them. Under standard conditions of temperature and pressure, the average spacing between molecules in steam will be 10 times the diameter of the molecule. Gas molecules exert practically no cohesive force upon one another. This lack of attracting force and their high-moving velocity explains why gases are void of shape and volume.

One of the more unusual characteristics of a gas is its unlimited capabilities of expansion. Regardless of the amount of gas placed in a container, it will always expand until the container is completely filled. If only half as much gas is placed in a container as what is needed to fill it, the container will still be full, but at a lower pressure. For this reason, it is nearly impossible to develop a complete vacuum. No matter how much air is pumped from a container, the remaining air will always redistribute itself throughout the container.

Kinetic Theory of Matter

The kinetic theory of matter is an attempt to explain how a substance behaves with respect to the properties of molecules that are used in its composition. In this regard, anything that moves does a certain amount of work and possesses some energy. The energy that a body has because of its motion is called kinetic energy. The moving molecules of matter each possess a discrete amount of energy. Changes in molecular energy or its energy level has a great deal to do with the state of matter.

When all additional form of energy, such as heat, is applied to a particular substance, it adds to the kinetic energy of moving molecules. This action tends to cause a decided increase in the velocity of each molecule. As a result of this, there is a corresponding increase in the temperature of a substance. When heat is transferred to another material, the energy level of each molecule is reduced accordingly. This action causes a decided reduction in molecule velocity and the internal temperature of the substance.

Changes of State

A state change in matter is brought about primarily by altering the energy level of individual molecules. The kinetic energy of each moving molecule is either increased or decreased according to the outside source of energy. In a sense, we can say that energy is the primary agent that brings about a change in the state of matter. Energy appears in many different forms, the most common of which are heat, light, electricity, magnetism, sound, mechanical action, nuclear energy, and chemical energy. Heat, chemical, and electrical energy are probably more responsible for most changes in matter associated with building equipment than all of the others.

When a solid piece of matter is heated, each molecule has a tendency to move more rapidly or have increased velocity. As a rule, a particular molecule does not move very far from its original position. At a given temperature, however, each type of matter encounters a rather unusual condition. A further increase in temperature does not cause a corresponding increase in molecule velocity. Instead of increased velocity, a solid will change into a liquid. The temperature at which solid matter changes into a liquid is called the melting point. Iron melts at 2800°F (1588°C), copper at 1083°F (634°C), and ice at 32°F (0°C).

Liquefaction

If a mixture of water and ice is heated gently, some of the ice will begin to melt but the temperature of the mixture will remain at 32°F

(0°C). The heat that is added is absorbed by the ice in melting. This illustrates, in effect, that a melting solid absorbs heat without causing a change in temperature. The kinetic theory of molecular energy helps in understanding this idea. When a solid is at its melting point, its molecules move so rapidly that their cohesion is not adequate to hold them together. Any additional heat added at this time will be received by the solid but will not cause an increase in molecular velocity. It will, however, reduce the cohesive force of the solid. As a result of this, ultimately the molecules begin to move more freely in the liquid. This means that the solid melts but that its temperature, which is based on molecular motion, remains the same. Only after the melting process is complete will the temperature begin to rise. The process of changing a solid into liquid is called *liquefaction*.

Solidification

The process of changing matter from a liquid state to a solid is called *solidification* and in some cases *freezing*. To solidify a liquid, heat must be removed from a material. In effect, heat moves only from a warm object to something of a lower temperature. When water freezes, it liberates heat to its surroundings. To freeze a liquid, it must be placed in an environment that is colder than the freezing point of the material. In effect, when a material loses heat its molecular velocity slows down.

To freeze a liquid when it is at its freezing point, heat must be transferred away or taken from the material. While a liquid is in the process of freezing, its temperature does not change. In a sense, we are removing heat from individual molecules, which causes them to begin to slow their movement. When they move slowly, there is greater cohesion between individual molecules, which causes them to be more reluctant to move. As a result of this, a material at its freezing point begins to solidify. Its temperature remains the same, however, until all of it has frozen. Only then will further cooling cause a change in temperature. In a sense, solidification is the reverse action of liquefaction.

Evaporation and Boiling

When water is placed on a floor during a cleaning operation, it tends to dry up very quickly after a short period of time. In this situation, we say that the water has *evaporated*. Essentially, this means that floor water has changed from a liquid state to a gas or vapor. Evaporation results when molecules at the surface of a liquid have enough kinetic energy to escape from the main body of the liquid.

When a liquid is heated, its molecules have a tendency to move faster and faster, so that more of them are able to pull away from its surface and escape into the air. With continued heating, a temperature is soon reached at which not only the surface molecules escape but those within the liquid gain also enough kinetic energy to escape as vapor. Bubbles of vapor within the liquid begin to form when boiling occurs. The temperature at which this takes place is called the *boiling point* of the liquid. Water boils at 212°F (100°C) and mercury boils at 675°F (357°C). When we think of boiling, we generally envision a material as being extremely hot. For many materials this is not necessarily true. A special hydrocarbon material, called chlorofluorocarbon gasses, boils at –21.6°F (–30°C).

Evaporation is quite different from boiling. Evaporation can take place to a greater or lesser degree at any temperature. Boiling is much more restricting and takes place at only one temperature for a specific material. Also, evaporation takes place only from the surface of a liquid, whereas boiling occurs throughout the liquid.

The rate at which a liquid evaporates depends a great deal on the nature of the material. Alcohol, chloroform, and ether evaporate much more rapidly than water. Regardless of the readiness of a material to evaporate, there are four ways in which evaporation may be increased.

1. *By adding heat.* When heat is added to a liquid, its molecules tend to become more active. This permits them to escape from the surface of the liquid rather easily, which increases evaporation.

2. *By spreading the liquid over a wider area.* When a liquid is spread over a large surface area, individual molecules tend to appear closer to the surface. This action gives the individual molecule an easier chance to escape, which improves the evaporation process.

3. *By decreasing the pressure upon the liquid.* A decrease in liquid pressure causes the air above a liquid to offer less opposition to each

individual molecule. A decrease in liquid pressure lets molecules move easily out of the liquid.

4. *By steadily replacing the moisture-laden air above the liquid with new air.* Fanning or blowing air above the surface of a liquid causes more molecules to move into the new air without having an opportunity to return to the liquid.

When water is heated, its temperature will rise rather steadily until it reaches 212°F (100°C). After this point, continued heating will cause the water to boil and change into steam. The temperature of the liquid, however, continues to remain at 212°F (100°C). This characteristic of water illustrates two rather important facts about boiling liquids. The first deals with the idea that the temperature of a boiling liquid remains the same as long as it continues to boil. Second, it takes a definite amount of heat to change a liquid into a gas or vapor state. According to the kinetic theory of energy, a specific amount of heat is needed to reduce the cohesive forces of a liquid so that its molecules may become separated from their neighbors.

Condensation

In the summer, a cool glass of water quickly gathers moisture on its outside surface when placed in a warm room or area. This accumulated moisture, which is technically called *condensate*, is generated from the air. *Condensation* is the resulting effect that occurs when a gas or vapor is changed into a liquid. This condition represents a very important molecular-state change.

When a small amount of vapor condenses, it has a tendency to release the heat that it absorbed when it went through the evaporation process. This is generally called the *heat of vaporization.* Specifically, when 1 gram of steam at 212°F (100°C) changes to water it gives off 540 calories of heat to its surroundings. This same amount of heat in the radiator of a steam heating system will release 540 calories of heat to a room when it condenses. Condensation is an essential function in the operation of an air-conditioning system and a heat pump. Hot gas flowing through the inside of metal pipes or tubes is changed into liquid by circulating cooler air or water over the outside surface of the pipe network. For condensation to occur, a gas or vapor must be subjected to a cooler environment.

HEAT

Heat has a tremendous influence on our lives. The amount of heat that is present has a great deal to do with the clothes we wear, the food we eat, and the temperature of our homes or buildings. In many localities we must increase the heat in homes and buildings in winter and reduce it in summer. We try to dress so that much of our body heat is retained in winter and given off in summer. Heat is also used to cook food to make it edible and is removed from food to preserve it.

Heat Energy

We know that all matter is composed of tiny particles called molecules. Most scientists believe that molecules are in a constant state of motion or vibration. They also believe that a body in motion possesses kinetic energy and that the faster it moves, the more kinetic energy it has. Moving molecules in matter represent kinetic energy that appears in the form of heat. The faster the molecules move, the more kinetic energy or heat there is in a particular substance. When the molecules of a substance slow down, there is less kinetic energy and substance heat.

Heat and Temperature

The terms *heat* and *temperature* are extremely important and must be clarified at this point. Heat is generally described as a measure of quantity, whereas temperature refers to a level of intensity. To distinguish more clearly between these terms, assume that a 5-gallon bucket and a 1-gallon container are filled with water of the same temperature. Water in the bucket will possess much more heat than the container because there are more molecules in it. Heat is directly dependent on both the speed and the number of molecules. A larger volume of water will therefore possess more heat than will a smaller volume. On the other hand, temperature is dependent primarily on the speed with which molecules move. The temperature of a body is therefore determined by the average kinetic energy developed by each molecule. Heat is quite different because it is determined by the total kinetic energy of the molecules. Temperature is measured in degrees with a thermometer, whereas heat is measured in British thermal units (Btu).

Heat Transmission

To make the best possible use of heat, it is important to know how it is transferred from one point to another, and how its movement can be stopped. In an apartment building, heat may be transferred from the basement to every room of the building in the winter months. In the summer, heat is transferred from each room of a building to an area outside or away from the building.

The transfer of heat between objects or areas is determined primarily by temperature. Heat must always flow from a warmer object to a colder object. The transfer rate of heat flow depends on the temperature difference between objects. Consider, for example, what would take place when two objects, one small and one large, are placed in an insulated box of suitable size. We must also assume that the temperature of the smaller object is twice that of the larger object. The heat level of the larger object is much greater than the smaller object, depending on the volume ratio of the two. However, because of the difference in temperature, heat will be transferred from the smaller object to the larger until they have both reached the same temperature. The length of time that heat flow occurs is dependent on the volume ratio of the objects.

The transmission of heat from one location to another depends a great deal on the material of the flow path. Heat will flow effectively through solids, gases, or liquids. The transmission of heat, however, can be achieved only by conduction, convection, or radiation.

If one end of a solid metal bar is placed in an open flame, the other end will soon become hot. The process by which heat is transferred from the flame to the cold end of the bar is called *conduction*. For conduction to take place, the heat source must touch the object being heated. Heat transfer by conduction is shown in Figure 2-1.

According to the kinetic theory of energy, conduction takes place by means of the collision of molecules. Heat energy from the source causes individual molecules at one end of the bar to begin to move more rapidly. Increased molecular velocity in this area causes collisions between adjoining molecules. As a result of this condition, these molecules also begin to move faster. The process continues until the iron molecules at the other end of the bar are reached. Heat transfer by conduction is dependent on the material and its physical size.

The conduction process applies primarily to heat transmission

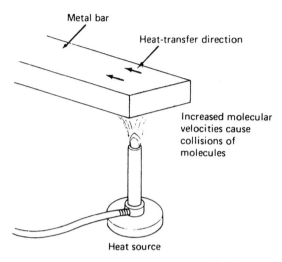

Figure 2-1. Heat transfer by conduction.

through solid materials. Metals are the best conductors of heat, whereas nonmetals in general are poor conductors or heat insulators. Cooking and food preparation rely very heavily on the transfer of heat by conduction.

Convection is the process of transmitting heat through a fluid such as a liquid or a gas. When a container of water is placed on a heat source, at first only the water at the very bottom receives heat by conduction through the container. Since most liquids are rather poor conductors of heat, very little heat is transferred to other parts of the liquid. As the bottom layer of water begins to receive heat, it tends to expand somewhat. This condition causes it to become less dense than the cooler water above it. As a result of this condition, warm water begins to rise to the surface, which causes heavier cold water to flow down toward the source of heat. The new bottom layer of water will receive heat, expand, and rise to the top of the container in the same manner as before. Through this process, each layer of liquid moves gradually downward, receives heat, and then rises to the top of the container. If the heat source is continued, the process is repeated over and over again until boiling occurs. Figure 2-2 shows an example of the convection principle applied to water heating.

Air or gas heating is an important application of the convection principle. Cool air near the surface of a heating element will derive

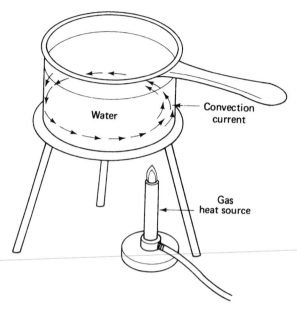

Figure 2-2. Convection heating of water.

heat energy initially by conduction. Heat transferred to molecules of air causes it to become less dense. As a result of this, warm air currents tend to rise and move away from the element surface. The warm air is replaced with cool dense air that normally flows down. A large part of the heat within a building is transferred by the convection principle.

Radiation is a process by which heat is transferred through the motion of waves. A prime example of this is the heat that reaches the earth from the sun. Since the space between earth and the sun is generally void of molecules, except near the earth's surface, its heat cannot be transferred by conduction or convection. In effect, heat energy is given off or radiated away from a heat source through infrared rays. These rays possess discrete bundles of energy that are set into motion by the source. Energy of this type moves away from the source in a wavelike pattern at the speed of light. An interesting characteristic of radiation is that the air between the heat source and the object, which the waves must pass through, is not heated. Figure 2-3 shows examples of heat transfer by radiation.

A stone dropped into a lake or pond sets up a disturbance that

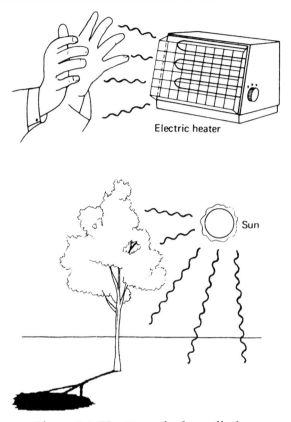

Figure 2-3. Heat transfer by radiation.

causes waves to occur on its surface. These waves start at the point where the stone first enters the water and move away in a ring pattern in all directions. The molecules of a heated object vibrate and produce waves that are very similar to these. The primary difference between water waves and heat waves is the medium in which the wave travels. Heat waves are unique in this regard, because they do not travel in matter. When a heat wave strikes an object in its path, heat energy is released into the object.

Heat Measurement

Two measures of heat are the calorie and the British thermal unit (Btu). A *calorie* is the amount of heat energy needed to raise the temperature of 1 gram of water by 1 degree on a Celsius (C) thermometer.

This unit is usually quite small and generally used only in scientific measurements. A *Btu* is an indication of the amount of heat needed to raise the temperature of 1 pound of water by 1 degree on a Fahrenheit (F) thermometer. Applications of Btu measurements are commonly used in heating systems, air-conditioner ratings, and water heaters. As a measure of heat energy, 1 Btu is equivalent to 252 calories.

Specific Heat

Each type of matter has a very unique characteristic with respect to its ability to absorb heat. The quantity of heat needed to raise the temperature of 1 gram of a particular substance by 1 degree Celsius is called the *specific heat* of that substance. As a rule, specific heat is a measure of the ability of a material to absorb heat compared with the ability of water to absorb heat. Water tends to absorb heat or lose it very slowly. The specific heat of water is 1 calorie, or simply the value 1. Compared with other materials this is a rather high value. Copper, for example, has a specific heat of 0.093. This means that it takes only 0.093 or about 1/10 calorie to raise the temperature of copper by 1°C.

Because of the high specific heat of water, regions bordering on large bodies of water tend to have cooler summers and warmer winters than do inland areas at the same latitude. In the summer the water absorbs large quantities of heat from the air and surrounding land. This causes a significant cooling action. In the winter, large bodies of water release heat into the immediate area, causing it to be warmer. The specific heat rating of materials used in air conditioners and heat pumps is changed to absorb and release heat to produce a desired condition.

Sensible Heat

Heat added to or removed from a particular substance that causes a change in temperature but not a change in state is called *sensible heat*. The word "sensible" refers to heat that can be sensed or measured by an instrument, such as a thermometer. Nearly all substances that are in a solid, liquid, or gaseous state possess some degree of sensible heat.

An example of sensible heat occurs in a water heater during its operational cycle. First, let us assume that the temperature of water in a storage tank is 140°F (60°C), as indicated by a thermometer (see

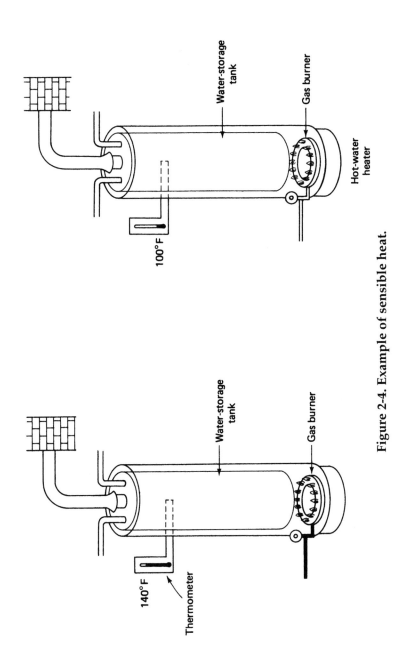

Figure 2-4. Example of sensible heat.

Figure 2-4). If several gallons of water is removed from the tank, its temperature may be lowered to 100°F (39°C). The temperature change in this case is representative of 40°F (21°C) of sensible heat. This pronounced change in tank temperature would normally be detected by a sensing device that would actuate the heat source. Applied heat causes the water temperature to rise from 100°F (39°C) to 140°F (60°C). This change in temperature sensed by the thermometer represents 40°F (21°C) of sensible heat. In an operational sequence of this type, a certain amount of time is required to cycle through the change in temperature.

Latent Heat

Heat added to or removed from a substance that brings about a state change but not a change in temperature is called *latent* or hidden *heat.* The graph of Figure 2-5 shows where latent heat occurs when water goes through a change in state.

Starting at –40°F (–40°C) on the graph, assume that heat is added to 1 pound of water in its solid state. Heat applied to ice will cause a corresponding increase in its temperature. This action is indicated by a rise in temperature between points 0 and *A* of the graph. Sensible heat is primarily responsible for this condition. Its value can be measured or observed on a thermometer. Approximately 36 Btu of heat is absorbed by ice in making this change in temperature.

When an additional 144 Btu of heat is added to the ice, as indicated between points A and B of the graph, the temperature remains at 32°F (0°C). The state change of melting ice absorbs a great deal of the applied heat. As a result of this, the temperature does not change in value. Latent heat, or the latent heat of fusion, is responsible for this condition. It continues until the melting process has been completed and the water sample is in its liquid state.

Point *B* to *C* of the graph shows where an additional 180 Btu of heat is added to water. This action causes a rise in temperature from 32°F (0°C) to 212°F (100°C). Sensible heat, which is indicated by a thermometer, is responsible for the change in temperature during this period.

Between points *C* and *D* of the graph, water is undergoing a state change from liquid to vapor. Here 970 Btu of heat is absorbed by the water without causing a change in temperature. Latent heat is involved in this process. The term *latent heat of vaporization* is generally used to describe this condition. The state change is complete at point *D* and the temperature would continue to rise beyond this point when additional heat is applied.

If the graph of Figure 2-5 is followed from point *D* through 0, it would show that the action just described is reversible. Essentially, this means that the same amount of heat that is absorbed by a substance must be released when the transfer process is reversed. Heat absorbed by a gas or vapor that is released when it is cooled is called the *latent heat of condensation*.

PRESSURE

Nearly all buildings today employ a system that involves some form of pressure in its operation. These systems perform a particular work function that is the result of pressure being transferred through a confined network of pipes or tubes. Air-conditioning equipment, heat pumps, freshwater systems, and fuel distribution are representative applications of pressure. The behavior of fluids and gases under pressure is an extremely important concept in building system operation.

Force and Pressure

The weight of any material has a very significant downward *force* that acts upon whatever is supporting it. The force developed by a material is measured in units of weight such as pounds, ounces, and grams. The total weight of the material is distributed over the surface area on which it lies. *Pressure* is defined as the intensity of a force that is applied to a specific unit area. Mathematically, this is expressed by the formula

$$\text{pressure} = \frac{\text{force}}{\text{area}} \quad \text{or} \quad p = \frac{f}{A}$$

If a 20-pound piece of metal rests on a surface of 5 square inches, there will be a pressure of 20/5, or 4 pounds of force, on each square inch of the surface. Pressure, as noted in this example, is measured

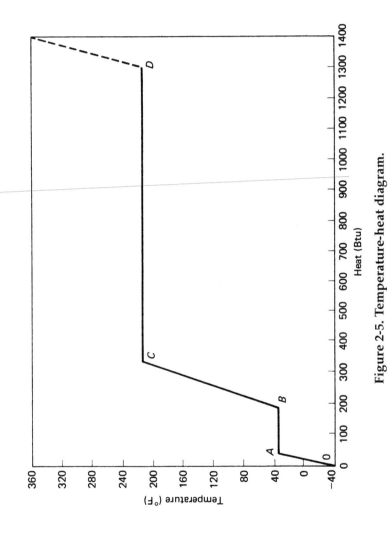

Figure 2-5. Temperature-heat diagram.

in pounds per square inch (psi) or pounds per square foot (lb/ft^2). Metric expressions are in grams per square centimeter (g/cm^2) or newtons per square meter (N/m^2). The name pascal (Pa) is also being used in some applications as a measure of pressure in metric values. One Pa equals 1 N/m^2. Being small, this measure may also be expressed as kilopascals (1000 Pa) or as megapascals (1,000,000 Pa).

Gas Pressure

A *gas* may be defined as a material in which there is very little cohesion between individual molecules. As a result of this, any material in a gaseous state tends to fill the entire space into which it is placed. The application of heat to a gas causes the velocity of individual molecules to increase quite significantly. As a result of this condition, individual molecules have a tendency to bombard the internal surface of the container. This action is representative of *gas pressure,* because molecular surface bombardment is a type of force against the wall of the container.

The relationship between gas pressure and temperature is a very important principle in air conditioner and heat pump operation. This relationship is expressed mathematically as

$$\frac{P_1}{T_1} = \frac{P_2}{T_2}$$

It means that the original pressure (P_1) divided by the original temperature (T_1) will equal the new pressure (P_2) divided by the new temperature (T_2). This relationship holds true as long as the volume of the gas is kept constant.

It can be concluded that the pressure exerted by a fixed volume of gas depends on the temperature of the gas. This principle is widely used in many temperature-measuring instruments.

Atmospheric Pressure

A gas pressure that is always present is exerted on the earth by the weight of the atmosphere above us. Because of the earth's gravitational pull, gases in the air tend to exert a continuous force on the surface of the earth. This force per unit area is called *atmospheric pressure.* Atmospheric pressure at sea level is 14.7 psi or 101.36 kPa. This value is normally used as a reference or a comparison with other pressure values. Atmospheric pressure decreases in value as the height above sea level increases. Atmospheric pressure is measured with an instrument known as a *barometer*. Barometric pressure values in specific localities are subject to a great deal of change due to variations in weather conditions.

Liquid Pressure

Matter in a liquid state has a rather limited attraction between its individual molecules. The force of attraction is such that if a liquid is poured into a container it will conform its shape to that of the container and fill it to a uniform level. For comparison purposes, a gas will completely fill a container and a solid will retain its original shape regardless of the container.

Liquids are fairly dense in their structure. In effect, this means that the force of gravity on a liquid is rather substantial. A 1-cubic foot (ft^3) volume of water weighs 62.4 pounds. The density of water is therefore 62.4 lb/ft^3. Mercury, by comparison, has a density of 846 lb/ft^3, and that of gasoline is 41.2 lb/ft^3. Scientifically, density is considered to be the weight per unit volume of a substance. It can be expressed in pounds per cubic inch (lb/in^3) or in pounds per cubic foot (lb/ft^3).

The weight of a fluid or liquid is somewhat different from that of a solid material. Liquids exert both a downward force on the bottom of the container and a lateral force on the sides of the container. If the container of Figure 2-6 is filled to 12 inches or 1 foot, we would have 1 cubic foot of water weighing approximately 62.4 pounds. The force exerted on the bottom of the container would be 62.4 lb. Expressed as pressure, this would be 62.4 lb/ft^2 or 0.433 lb/in^2. A square foot is 12 × 12 in, or 144 in^2. Therefore, 62.4/144 equals 0.433 psi. The same pressure would also appear at the sides on the bottom of the container. Fluid pressure is the same on each square inch of the walls of the container at the same depth. This pressure acts at right angles to that at the bottom of the container.

Head Pressure

Pressure and the depth of liquid in a container have a very close relationship in fluid applications. Pressure is developed as a result of

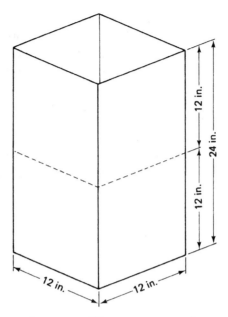

Figure 2-6. Water storage tank.

a fluid being elevated to a position or "head" well above other components. The depth of a body of water also has *head pressure* anywhere below its surface. Fluid pressure is directly dependent on the depth or height of liquid in a container.

If the tank of Figure 2-6 is filled to the top, it would be capable of holding 2 cubic feet of water. The increased weight of the water in this example would now be 2 × 62.4, or 124.8 lb. The force exerted on the bottom of the tank would also be 124.8 lb, distributed over the same area of 144 square inches. Head pressure at the bottom of the tank would now be 124.8/144, or 0.866 psi. This represents twice the pressure that the tank exhibited when it was filled to a depth of 1 foot. Pressure is, therefore, based upon the height or depth of the head. Actual values can be determined by multiplying the head by the weight of the liquid. For an open container of water, the pressure would be 0.433 × the head in feet.

Absolute Pressure

A pressure value measured from a perfect vacuum is called *absolute pressure*. In effect, atmospheric pressure and absolute pressure have the same meaning. Absolute pressure is generally expressed in pounds per square inch absolute, or psia. A common way of determining absolute pressure is to add 14.7 psi to a value indicated by a pressure gauge.

Gauge Pressure

A pressure value of zero pounds per square inch gauge, or psig, is indicated by an instrument when it is not connected to a pressure source. Pressure values below 0 psig are considered to be a negative value or inches of vacuum. Special instruments called *compound gauges* are used to measure pressures both above and below atmospheric pressure. These gauges employ two or more pressure scales. One scale measures values above atmospheric pressure while the other determines levels below atmospheric pressure. Figure 2-7 shows a compound pressure gauge that is used to indicate pressure above the

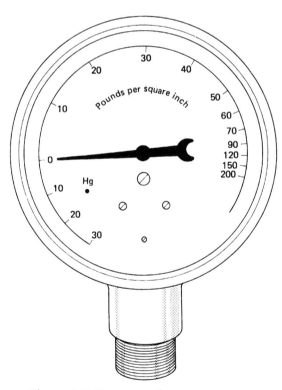

Figure 2-7. Compound pressure gauge.

atmospheric value in psi and below atmospheric values in inches of mercury (Hg).

Pressure and Temperature

The *pressure-temperature relationship* of fluids and gases is extremely important in some building equipment. The point at which a liquid will change into a gas is directly dependent on its applied pressure and temperature. For each pressure applied to a liquid, there is a specific temperature value at which it will begin to boil and cause a state change. In effect, this is based on the liquid not being contained by another material.

In practice, since all liquids react in essentially the same way, pressure often provides a convenient way of regulating. temperature in a confined area or space. Liquid circulating inside a pipe network that is isolated from the atmosphere can have a pressure value that will cause boiling or evaporation to occur when its outside is exposed to heat from a room or space. This action will permit heat to be carried away from a space by a change in liquid temperature. Cooling is achieved through the evaporator coil of an air conditioner by this process.

The pressure-temperature relationship is reversible. When the pressure of a gas inside a closed pipe network is increased to a point at which it will cause gas temperature to become higher than the surrounding air, heat will be released. This, in effect, will cause the gas to condense or change into a liquid. The condenser of an air-conditioning system operates on this principle.

HUMIDITY

The amount of water vapor or moisture content of air is commonly described by the term *humidity*. When air holds a great deal of moisture, its humidity level is considered to be high. If there is very little moisture in the air, its humidity level is low. *Saturation* occurs when air holds as much moisture as possible at a specific temperature and pressure. More moisture is required to saturate air when the temperature is high than when it is low. As a general rule, the air of a building can usually hold additional moisture because it rarely, if ever, reaches saturation.

The humidity level of air has a great deal to do with human comfort. Building air that is too dry tends to irritate a person's nasal passages, lips, eyes, and skin. The air in a building that has too much moisture feels damp and heavy. This condition causes the body perspiration rate to slow down and may aggravate respiratory difficulties. Proper control of the air moisture in a building improves human comfort and increases the operational efficiency of heating and air-conditioning equipment.

One measure of air moisture is described as *specific humidity*. This term refers to the actual weight of moisture contained in a given amount of air. The expression "grains of water per pound of dry air" is a common measure of specific humidity. A grain (gr) is equal to 1/7000 pound. Air at 50°F (10°C) is saturated when it contains 53.5 grains per pound. At 80°F (27°C), saturation occurs when the air contains 155.8 gr/lb.

By experimentation, it was found that human comfort is related much more closely to *relative humidity* than to specific humidity. Relative humidity (RH) refers to the amount of water vapor in the air compared to the amount needed to achieve saturation. If air contains only half of the amount of moisture that it can hold when saturated, the RH is 50%. Ideal inside building conditions during the winter occur when the temperature is 68°F (20°C) at a relative humidity of 50 to 60%. Below this level, perspiration evaporates rapidly from our skin, which causes an uncomfortable cool feeling. When the RH level is higher than this figure, perspiration evaporation is reduced and our skin becomes uncomfortably moist and warm. In warm climates and summer air-conditioning applications, an ideal figure of comfort is 78°F (26°C) at 60% RH. The effects of humidification have a tremendous influence on the operational efficiency of heating and cooling equipment.

DEW POINT

The dew point is actually a temperature that can be measured on the Fahrenheit or Celsius scales. The dew point is the temperature at which moisture (water) will precipitate (condense into water, such as rain in a cloud), just as dew is found on the morning grass after a cool overnight. Often this temperature is cooler than ambient temperature because air must be cooled in order for the water vapor to condense. It is the dew point that determines the relative humidity previously discussed. High relative humidity indicates the dew point is near the ambient temperature. Low relative humidity indicates the dew point is much cooler than the ambient temperature. Should the dew point temperature occur below freezing, it is termed the frost point.

WORK

Scientifically, *work* is accomplished when an applied force moves a body or mass through a measurable distance. In this definition the term "force" refers to a push or pull that tends to modify motion, and "distance" refers to a positional change in object location. Force is normally measured in units of weight, such as pounds, ounces, or grams. Distance, by comparison, is a linear expression that is measured in inches, feet, or centimeters. The resulting work that is accomplished is described in foot-pounds, ounce-inches, and gram-centimeters. When a force actually succeeds in moving a body, we say that a specific amount of work has been accomplished. Mathematically, work is expressed by the formula

$$\text{work} = \text{force} \times \text{distance} \quad \text{or} \quad W = FD$$

As an example, let us assume that a person moves a 40-1b (18-kg) box 10 ft (3.1 m). The amount of work done is

$$W = FD \qquad\qquad\qquad W = FD$$
$$= 20 \text{ lb} \times 10 \text{ ft} \qquad\qquad = 18 \text{ kg} \times 3.1 \text{ m}$$
$$= 200 \text{ foot-pounds (ft-lb)} \qquad = 55.8 \text{ kilogram-meters (kg-m)}$$

For work to be accomplished, some outside agent, such as a person, animal, steam, electricity, or wind, must be employed. This agent is primarily responsible for developing the force that causes motion. In practice, any motion that occurs is produced against some form of resistance. A person pushing a box across the floor produces motion that is done against the resistance of the floor.

Power

A more realistic concept of work must take into account the length of time that a force is acting on an object that is being moved. The term power is commonly used to express this relationship. Technically, power refers to the time rate of doing work. In this regard, power involves force, distance, and time. It is determined by dividing the amount of work accomplished by the expended time. Mathematically, power is expressed by the formula

$$\text{power} = \frac{\text{work}}{\text{time}} \quad \text{or} \quad P = \frac{W}{T}$$

An example of this occurs when a machine lifts a 100-lb (44.8-kg) weight a distance of 10 ft (3.1 m) in 10 seconds. The power involved in this operation is

$$P = \frac{W}{T} \qquad\qquad P = \frac{W}{T}$$

$$= \frac{100 \text{ lb} \times 10 \text{ ft}}{10 \text{ sec}} \qquad = \frac{44 \text{ kg} \times 3.1 \text{ m}}{10 \text{ sec}}$$

$$= \frac{1000 \text{ ft-lb}}{10} \qquad = \frac{136.4 \text{ kg-m}}{10 \text{ sec}}$$

$$= 100 \text{ ft-lb/sec} \qquad = 13.64 \text{ kg-m/sec}$$

Horsepower

The power used by most building equipment today is expressed in a more practical unit called *horsepower* (hp). Originally, this referred to the average amount of work that a draft horse could perform. James Watt of England concluded that a steady 150-lb force could be applied by a horse while walking 2-1/2 miles an hour. The horse thus performed work at a rate of 33,000 ft-lb per minute or 550 ft-lb per second. The term "horsepower" (in English units) is now commonly used to express a measure of mechanical power. A piece of equipment with a rating of 1 hp can do 550 ft-lb of work during every second that it is in operation. A 2-hp machine can work twice as fast as a 1-hp machine. Electric motors are commonly rated in horsepower and fractional horsepower values.

ENERGY

Matter has the capacity to do work under certain conditions even though it may not be accomplished at a particular moment. A moving stream of water or water stored behind a dam, for example, can be made to do work in a variety of applications. Gas or air compressed in a tank can also do work when it is released at a particular time. This capacity of matter to do work is generally called *energy*.

Importance of Energy

All life on earth depends in some way on energy. Most of the earth's energy comes from the sun. It travels to the earth in rays or waves. Energy in this form is used by plants to make food. Plant food is needed to sustain life by providing nourishment for our bodies and muscles. The sun's energy is also stored in coal, wood, and oil, which is also used to produce food and modify matter. It has been estimated that the amount of energy falling on the earth's surface annually is equal to the energy of 250 million-million (tera) tons of coal.

Energy is considered to be one of the two fundamental essentials of physical science. The other part of this is matter. These two items are not generally considered to be completely independent. Many physicists believe that matter and energy are, in fact, two different aspects of the same thing. This would be comparable to the state of water being either liquid, solid, or vapor. They also believe that the electrons of an atom give off or release energy in the form of light. A body loses some of its mass when energy is released and gains mass when energy is absorbed.

Sources of Energy

Energy is known to exist in a variety of different forms. Among these are heat, light, chemical, electrical, sound, mechanical, and nuclear. As a general rule, energy changes rather easily from one form to another. The steam-driven turbine generator of a power plant is designed to convert heat energy into electrical energy. The heating system of a building operates by converting the chemical energy of gas or fuel into a usable form of heat. Electrical energy may also be converted into light, heat, or mechanical energy through the use of different equipment. The capacity of energy to do work makes all of this possible.

Classification of Energy

All forms of energy are divided into two unique classifications which deal with either motion or position. *Potential energy* is used to describe energy of position or molecular arrangement. Energy of motion is commonly called *kinetic energy.*

A body containing potential energy is primarily due to its position, condition, or chemical state. In a sense, potential energy represents a form of work that has been done previously. A large box raised to a shelf is elevated with respect to a chosen energy level such as the surface of the earth. The amount of work expended in lifting this mass a certain height is changed into potential energy due to its position. A wound clock spring possesses potential energy because it can supply the work necessary to make a clock operate for many hours. Energy stored in a wound clock spring is the result of previous work that causes a change in its physical condition. The potential energy of gas or fuel oil is used to heat a building. This type of energy is based on the chemical state of matter. Burning a gas causes chemical energy to be released as heat energy.

Kinetic energy is based upon the movement of a particular body. Every body placed in motion does work and continues to do work until it reaches zero velocity. If an elastic band is stretched between two fingers, it represents potential energy. Releasing the band causes it to snap back to its normal position. Potential energy is changed immediately to kinetic energy as long as the band is in motion. The word "kinetic" comes from a Greek word that means to move. A moving motor shaft, circulating hot water, and electron flow in a conductor all exhibit kinetic energy.

Conservation of Energy

If a given amount of energy is followed through various changes, we find that its total energy does not essentially change. The law of conservation of energy states that the amount of energy in the universe is always the same. It is possible to get only as much energy out of a machine as you put into it. In practice, output energy in a usable form is always somewhat less than the input. A portion of useful energy is usually wasted as heat in overcoming friction. Losses of this type prevent machinery from operating endlessly or from having perpetual motion. In effect, energy is neither increased nor decreased but can only be changed into another form.

The pendulum of Figure 2-8 demonstrates the law of *conservation of energy*. When the weight or bob of the pendulum is moved from its resting position at point *B* to point *A*, work is being done. At point *A* the bob has attained a certain amount of potential energy due to its new position. Releasing the bob causes it to swing back to position *B* and on to position *C*. The potential energy stored at point *A is* transformed into kinetic energy as the bob passes point *B*. Upon reaching point *C* kinetic energy is transformed momentarily into potential energy again. This, in turn, is again transformed into kinetic energy as the bob moves from *C* to *B* to *A*. The pendulum would, in effect, swing back and forth forever if it were not for energy lost through friction and the resistance of air. Overcoming these losses causes some energy to be transformed into heat. In a sense, the initial applied energy does

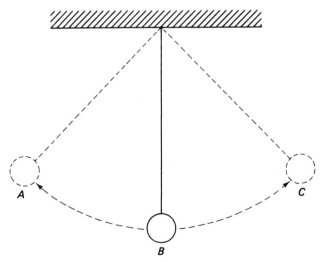

Figure 2-8. Conservation-of-energy-law demonstrator.

not change in value but appears in a different form.

Careful study of the previous example reveals that, when energy of one form disappears, other forms of energy appear in equal amounts. The operational equipment of a building is directly related to this law. In effect, we are very concerned with operation being achieved with a minimum of energy loss. Efficient equipment operation with reduced energy losses will help to keep the world's supply of useful energy in proper balance.

Chapter 3

The Building Structure/ Saving Money

INTRODUCTION

The first major system of a building that we consider is the building structure. The building structure includes walls, floors, ceilings, and external parts of a building. Commonly these parts are referred to as the building envelope.

The building envelope may not seem to be that important or may not appear to be a system. But, the way a building is pitted against the elements has a direct effect on the operation of other systems in the building and can result in very definite financial and environmental savings.

HEAT LOSS

Heat loss is a very important factor in the operation of a building. The amount of heat that is lost from a building into the outside air affects the amount of heating, cooling, and ventilation required in a building. Heat loss is measured in British thermal units (Btu). A Btu is the amount of heat needed to raise the temperature of 1 pound of water by 1 degree Fahrenheit (1°F). The amount of Btu of heat loss depends on the type of building structure, including doors, windows, walls, floors, and ceilings. The climate of an area is also important in determining heat loss.

Determining Heat Loss

Heat loss in a building is determined after identifying places where heat loss occurs. Because heat flows from a warmer area to a cooler area, it passes through doors, windows, walls, floors, and ceilings. The structure of each room area must be known to find the heat loss for each room of a building.

Heat loss that occurs through walls, floors, and ceilings is referred to as *transmission heat loss.* In all buildings, there is some cold air that comes inside. An equal amount of hot air goes outside. This process is called *infiltration heat loss.* Most infiltration usually takes place around doors and windows. Whenever there are cracks or any types of openings around windows or doors, air will leak into a building. Air leakage is caused by wind and the difference in inside and outside temperatures. As wind speed increases or a temperature difference increases, air leakage will also increase.

Measuring Heat Loss

Heat loss through parts of a building can be estimated by using measuring instruments. One such instrument is shown in Figure 3-1. This is an infrared temperature detector which measures heat without being in contact with a surface. The meter is aimed at the area where a temperature measurement is desired and the reading then appears

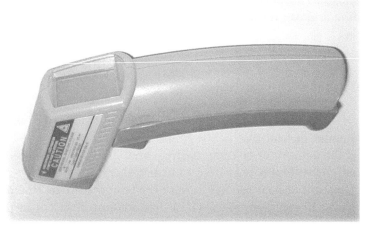

Figure 3-1. Infrared thermometer for determining heat loss. (*Courtesy of Mikron Instruments Co.***)**

on the digital readout of the meter. In this way, problem areas where excessive heat loss occurs can be identified. Additional information regarding temperature measurement can be found in Chapter 10, Instrumentation and Measurement.

Air Films and Spaces

On the outside of buildings and the inside of rooms, there is a movement of air over the surfaces of the structure. This film of air on the outside and inside opposes heat flow through the structure. The thermal resistance (R) value of the air film depends on whether it is inside or outside. The R-values of inside air films are based on (1) the position of the wall, floor, or ceiling (vertical, horizontal, or sloped); (2) the direction of heat flow; and (3) the *emittance* of the surface. Emittance is determined by how reflective the wall finish is.

Outside-air-film R-values are based on moving air rather than on the still-air values used for inside films. There are different R-values for winter and summer.

Air spaces in building structures are also used to resist the flow of heat. A space between materials is often left to increase the R-value of the material. R-values for structural materials are based on the same factors as those for inside air films: position of air space, direction of heat flow, and emittance.

Unheated Spaces—Heat Loss

Heat loss will occur from a heated room into an unheated space or room. The amount of heat loss depends on the difference in temperature between the heated room and the unheated space. The temperature of the unheated space will be higher than the outside temperature and lower than the inside temperature. Examples of unheated spaces are inside storage areas, uninsulated attic areas, and areas below floors.

Reducing Heat Loss

Heat loss may be greatly reduced just by sealing cracks between materials. Sealing cracks causes air infiltration to be reduced. Commonly, air infiltration occurs between the outside frame of a window or door and the wall around it. Expansion and contraction of buildings causes the cracks to become larger as the building gets older. A flexible sealant should be used to seal these cracks. Caulking should *not* be used because it tends to become brittle and break. It is easy to find areas where heat loss occurs on a cold day. On a cold, windy day the infiltration of air can be felt by placing your hand near any opening.

Heat loss also occurs around exhaust fans or ventilator units, electrical distribution panels, openings in storage areas of attics or basements, and in many other areas of industrial and commercial buildings. It is easy to make a quick inspection to identify areas of high heat loss and take corrective action.

To reduce heat loss it is essential to insulate properly all areas of a building that are exposed to colder temperatures. Heat loss can be greatly reduced in most buildings by adding insulation in areas such as crawl spaces and ceilings.

HEAT GAIN

The heat that passes into a building during warm weather is referred to as *heat gain*. Heat-gain estimations are necessary to determine the type of cooling unit that is needed for a building. Heat gain is essentially the opposite of heat loss. The same factors, transmission and infiltration, are used to determine heat gain. Transmission heat gain occurs through walls, floors, and ceilings and infiltration takes place through windows and doors. The total heat gain that occurs in a building is used to find the total *cooling load* of the building.

A major difference between heat loss and heat gain is that most heat loss usually occurs at night, and the greatest heat gain usually occurs during the day. Air infiltration into buildings in the summer is ordinarily much less than in the winter. This is due to the smaller temperature difference between inside and outside air and the lower wind speed in the summer in most areas.

Heat-gain calculations for buildings must also consider the heat given off by people and appliances. The number of people in a building, the number of appliances, and the amount of time these appliances are used will vary and are usually unpredictable. For these reasons, the heat gain they cause in a building must be estimated rather than calculated more precisely.

Solar radiation is an important variable to consider when calculating heat gain. Most solar radiation occurs through windows. Therefore, the orientation of the window (direction it faces), the kind of glass used, and whether or not shade trees or roof overhang are in front of the window affect solar radiation. The amount of shaded area of a window varies according to orientation, amount of overhang or permanently shaded area, and the geographic latitude of the building. Ordinarily, windows oriented toward the northeast or northwest cannot be protected by roof overhangs or shade trees. They are considered to be in maximum sunlight and minimum shade. Heat gain of buildings must be estimated to determine the cooling load of the building.

ENERGY USE IN BUILDINGS

There are several different uses of energy in buildings. The major uses are for lighting, heating, cooling, power delivery to equipment and appliances, and domestic hot water. The amount that each contributes to the total energy use varies according to the climate, type of building, and time of year. Energy use for heating is the largest area of use on a national level. In areas where very mild winters occur, cooling load will be greater than heating load in terms of the total energy use. In some types of buildings in certain climatic zones, the lighting loan might be greater than either the heating or cooling loads.

Industrial and commercial buildings are very similar in terms of energy use except that many industries use large quantities of energy for specialized processes. Most residential buildings use the greatest amount of energy for heating and cooling. It is difficult to generalize about energy use by type of building because there are so many variables that determine the energy use in a particular location. The best way to determine energy use is to analyze utility bills over a period of time as part of an *energy audit* or energy assessment of a building. This method is discussed in Chapter 11.

FACTORS THAT AFFECT BUILDING CONSTRUCTION

Energy conservation should be a major consideration in the design of all buildings. Many factors affect the type of building design used for buildings. The available building site, climate, budget, and the building function are among these factors. In retrofit or remodeling of older buildings, the same factors must be kept in mind.

Effect of Climate

The United States has many different climatic conditions. In terms of degree-days (see "Important Terms" in this chapter) the range is from 0 degree-days in parts of Hawaii to over 10,000 degree-days in areas of Alaska. Such conditions as swamps, deserts, forests, and mountainous areas greatly affect building design. Each building location has different characteristics in terms of temperature range, wind, humidity, and amount of sunlight.

The amount of sunlight has a tremendous impact on building design for energy conservation. It is possible to manipulate sunlight to accomplish wise energy use in a building. Some of the ways in which sunlight can be used to conserve energy are discussed in Chapter 9: Examples of using sunlight for energy conservation are solar collectors for heating or cooling and domestic hot-water systems or heat pumps assisted by solar collection methods. The amount of sunlight also affects the color that should be chosen for walls and roofs. A dark color should be used to absorb heat in north-facing areas in cold climates, whereas a light color should be used for reflection in warm climates.

Another important climatic factor in building design is the amount of wind and wind speed. Wind can increase the heating and cooling loads of buildings. Wind can also affect the humidity of a building. In northern areas, the north and west sides of a building are subjected to cold winds. Building entrances and areas with several window openings should not be oriented to the north or west in these locations to reduce heat loss.

Climate should be a major consideration in energy-conscious building design. Even factors such as the amount of rain and snowfall and the types of trees that grow in an area affect building design.

Obtaining Climate Information. It is possible to make energy-saving changes in a building based on local climate conditions. Local weather summaries are available from the Environmental Data Service, a government agency. The National Weather Service also provides monthly and annual weather summaries. These summaries contain such data as temperature averages, heating and cooling degree-days, precipitation, sunlight data, cloud cover, and wind data. These data are prepared and listed for most major cities in the United States. Many climatic factors must be considered when attempting to make energy-saving changes for an existing building or designing new buildings for maximum energy conservation.

Practical Considerations

There are many practical considerations that should be taken into account in the design of a new building or a retrofit of an older one. A few of these will be mentioned to stimulate thinking in regard to energy conservation.

The orientation of the building affects energy use. For instance, a roof that slopes toward the south is exposed to more sunlight than one facing to the north. Winds have more effect on energy use when openings in the building face north or west. Orientation can also affect the benefit that can be gained from natural lighting. Natural lighting is particularly helpful in terms of lower energy use in warmer climatic areas; however, heat loss through the areas that provide the natural light should be considered.

The use of round or spherical buildings, which have less external surface area, is a possibility. They allow architectural creativity in design and provide less heat loss or gain. Tall buildings provide less exposure to sunlight on the roof, but subjects the building to more wind. Overhangs can also be an important part of energy-conscious building design.

The floor plan of a building should reflect much thought about energy conservation. For instance, in the north section of a building, areas of low use by people, such as corridors, bathrooms, and storage areas, can be used as "buffers" between outside walls and interior rooms. Placing rooms that have similar heating and cooling needs together on the floor plan will simplify HVAC systems and thus reduce energy consumption. Conservation can also be accomplished by reducing the physical size of the building. Placing bathrooms adjacent to one another or directly above one another on multistory buildings will simplify ventilation and plumbing systems and causes less energy use. The reduction of windows and doors in a building will also reduce energy consumption by a reduction of heat loss or gain. Floor plan design has a tremendous impact on energy use in a building. Architects and designers should keep energy management in mind throughout the entire planning process for a new building or remodeling of an older building.

WINDOWS

The use of windows in a building structure plays an important role in the overall energy consumption of the building. Typically, there are far too many windows in a building. Of course, windows provide natural light, ventilation, and aesthetic benefits, but they can also cause discomfort to people near them due to excess cold, heat, or glare. Windows should be carefully placed on new buildings. There are also some things that can be done to help conserve energy by working with existing windows in buildings. Some of these techniques are discussed later in the chapter.

Types of Windows
There are many types of windows used in building construction. We look next at a few of the more popular kinds of windows. You should pay particular attention to the construction features of each type of window.

Casement Windows. A casement window is shown in Figure 3-2(a). This type of window has a crank that is rotated to cause the window to open. The window portion is mounted onto its frame with hinges. A cutaway view of a casement window with insulating double-pane glass is shown in Figure 3-2(b). The use of insulating glass is very important for energy conservation.

Awning Windows. An awning window is shown in Figure 3-3(a) and a cutaway is shown in Figure 3-3(b). This window is opened by pulling forward and allowing the window to slide along a grooved track on the sides of its frame.

Double-Hung Windows. The double-hung window, shown in Figure 3-4(a), is a very popular type of window, particularly for small

(a)

(b)

Figure 3-2. (a) Casement window with insulating glass; (b) cutaway view. (*Courtesy of Andersen Corp.***)**

(a)

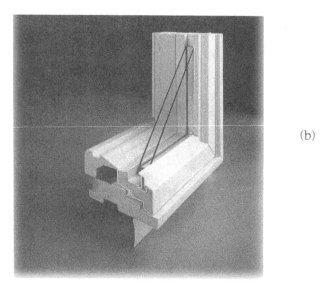

(b)

Figure 3-3. (a) Awning windows with double-pane insulating glass; (b) cutaway view. (*Courtesy of Andersen Corp.*)

(a)

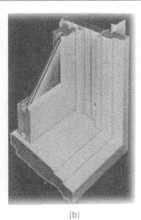

(b)

Figure 3-4. (a) Double-hung window with insulating glass; (b) cutaway view. (*Courtesy of Andersen Corp.*)

buildings. One section of window is mounted in tracks above and parallel to the lower section. Usually, either window will move up or down along the set of tracks. When closed, the window frames along the center of the window opening rest adjacent to each other to reduce heat loss. A cutaway view of an insulating glass double-hung window is shown in Figure 3-4(b).

Gliding Windows. Gliding windows, shown in Figure 3-5(a), slide along a track similar to that of sliding glass doors. A cutaway view of this type of window with insulating glass is shown in Figure 3-5(b).

Sliding Glass Doors. Sliding glass doors are used in many residential buildings and to a limited extent in some commercial and industrial buildings. They should be constructed with insulating glass to provide energy saving. A typical sliding glass door is shown in Figure 3-6(a). A cutaway view showing the snug-fitting design and insulating double-pane glass is shown in Figure 3-6(b).

(a)

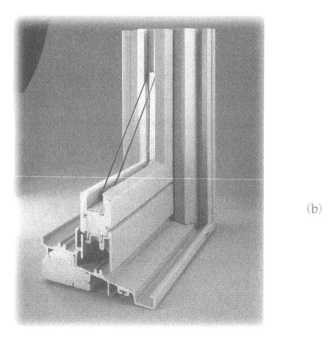

(b)

Figure 3-5. (a) Gliding window with double-pane insulating glass; (b) cutaway view. (*Courtesy of Andersen Corp.*)

(a)

(b)

Figure 3-6. (a) Sliding glass door with insulating glass; (b) cutaway view. (*Courtesy of Andersen Corp.*)

Storm Windows and Doors. For some types of buildings, installing storm windows and doors will provide a means of saving a great deal of energy. As much as eight times as much heat loss occurs at window areas as at a similar area of wall section. Storm windows and doors are placed on the outside of existing window and door openings. They come in a variety of sizes and styles to fit the needs of the building.

CARPETING

Carpeting can be used to beautify floors and also to provide energy savings. Carpet also reduces noise and the likelihood of accidents. Wall-to-wall carpeting on floors with pads placed underneath provides an air barrier which reduces heat loss through floors.

INSULATION

Most existing buildings were built when energy costs were cheaper than they are today. Most of these buildings do not have enough insulation; therefore, energy is being wasted. Insulation can easily be added to many areas of existing buildings. A typical estimate is that adequate insulation placed in an underinsulated building can save from 10 to 20% on energy costs. Of course, the actual savings depend on how much insulation the building had previously, how much was added, and other factors, such as the number of windows and doors and the quality of weatherstripping.

Types of Insulation

The types of insulation available include (1) mineral wool, (2) cellulose, (3) reflective foils, (4) vermiculite, and (5) various foam insulations. Several different types of insulation are shown in Figure 3-7.

A widely used type of insulation is mineral wool, which is either fiber glass or rock wool. Mineral wool insulation is available in several types, including blankets, blown insulation, poured insulation, and batts. Blankets come in rolls, with or without vapor barriers. Blankets

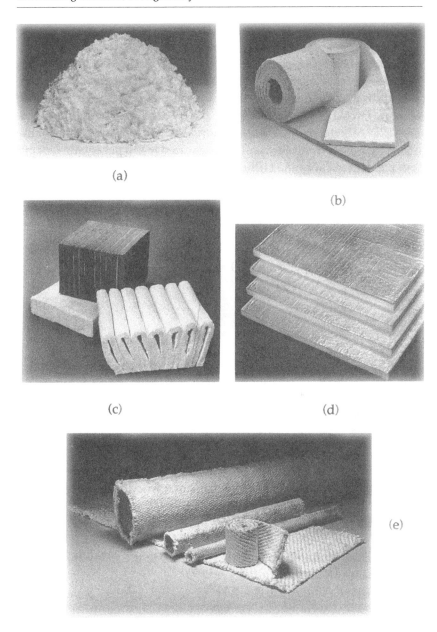

Figure 3-7. Types of insulation: (a) loose ceramic fiber; (b) ceramic fiber blanket; (c) modular; (d) industrial board; (e) textile. (a-c, e, courtesy of Carborundum Co.; d, courtesy of Certain Teed Corp.)

without vapor barrier are called unfaced insulation. Blown insulation is composed of loose pieces of insulation which are blown by air pressure into attics. Batts are like blankets but are precut to 4-foot or 8-foot lengths.

Insulation Considerations—R and U

The proper insulation of a building is a very important factor. Insulation is used to oppose the escape of heat. The quality of insulation is expressed by a *thermal resistance factor* (R). The total thermal resistance of a building is found by considering the thermal resistance of the entire structure (wood, concrete, insulation, etc.). The inverse of thermal resistance (I/R) called the coefficient of *heat transfer* (U), is an expression of the amount of heat flow through an area expressed in Btu per square foot per hour per degree Fahrenheit (Btu/ft^2/hr/°F). The following formulas are used in the conversion of either R or U to electrical units (watts):

$$\text{thermal resistance } (R) = \frac{1}{\text{coefficient of heat transfer}} = \frac{1}{U}$$

$$\text{watts } (W) = \frac{\text{coefficient of heat transfer}}{1} = \frac{U}{3.4}$$

or

$$\text{watts } (W) = 0.29 \times U$$

The manufacturers of insulation can supply various tables that can be used to estimate the heat loss that can occur in buildings of various types of construction. A building must have sufficient insulation to reduce heat loss; otherwise, electrical heating and air-conditioning systems will be very inefficient. The heat loss of a building depends primarily upon the basic building construction and the amount of insulation used. For instance, buildings made of concrete have a different amount of heat loss than those made of a wood frame construction.

The following sample problem will help you to understand the importance of adding insulation to a building.

1.   Given—A building constructed to provide the following thermal resistance (R) factors:

a. Exterior shingles are $R = 0.90$
b. Plywood sheathing is $R = 0.85$
c. Building paper used is $R = 0.05$
d. Wall structure has an $R = 0.90$
e. Wall plaster has an $R = 0.40$
f. Insulation is $R = 13.0$

2. Find—The total thermal resistance (R), the coefficient of heat transfer (U), and the watts (W) of heat loss both with and without the insulation.

3. Solution (without insulation):

$$R = a + b + c + d + e$$
$$= 0.90 + 0.85 + 0.05 + 0.90 + 0.40$$
$$= 3.10$$

$$U = \frac{1}{R}$$
$$= \frac{1}{3.10}$$
$$= 0.32 \text{ Btu}/\text{ft}^2/\text{hr}/°\text{F}$$

$$W = \frac{U}{3.4}$$
$$= \frac{0.32}{3.4}$$
$$= 0.094 \text{ watt heat loss}$$

4. Solution (with insulation):

$$R = a + b + c + d + e + f$$
$$= 0.90 + 0.85 + 0.05 + 0.90 + 0.40 + 13.0$$
$$= 16.1$$

$$U = \frac{1}{R}$$

$$= \frac{1}{16.1}$$

$$= 0.062 \text{ Btu}/\text{ft}^2/\text{hr}/°\text{F}$$

$$W = \frac{U}{3.4}$$

$$= \frac{0.062}{3.4}$$

$$= 0.01823 \text{ watt heat loss}$$

You can see from the results of this problem that adding insulation into the walls of a building has a great effect upon heat loss. The insulation has a much greater effect in controlling heat loss than do the construction materials used for the building.

Installing Insulation

Insulation should be used on all exterior walls and on interior walls between any heated or cooled area and unheated areas such as storage spaces. Ceilings that have cold spaces above should also be insulated. All floor areas that are above cold spaces should have adequate insulation.

The *R*-value (thermal resistance) of insulation is very important. R-values provide a means of knowing how well insulation resists the flow of air through it. Thickness of insulation affects the *R*-value; however, some insulations with the same *R*-value may be of different thicknesses. Typically R19 mineral wool batts are about 6 inches thick and R-13 blankets are about 3-1/2 inches thick. The R value of insulation is usually marked so that it is easily seen on the insulation or its package.

The recommended R-value for buildings, both new and existing, has steadily been increasing over the years. Some general estimates of R-value needed in certain areas a building for energy savings are.

In this example, the lower R-values are for existing structures, the higher for newer structures. These values will vary according to local climate, how the building is framed, and possibly to local building codes. In colder climates, it may be economically feasible to use materials that surpass these recommended values. In contrast, the values in 2007 were (1) attics, R49-R60; (2) floors, R25-R30; and (3) walls, R13-R21, and in 1993 they were (1) attics, R22; (2) floors, R11-R19, and (3) walls, R11. Local building contractors or insulation suppliers can provide information about the type and amount of insulation needed for a particular area.

Additional insulation can easily be placed in most buildings. In many cases, insulation can be installed without hiring a contractor. If attic insulation is added to attic floors where there is no insulation, batt insulation can be added between ceiling joists. Vapor barriers *must be placed face down* and the insulation does not have to be stapled. Adding insulation to areas such as uninsulated attics is a tremendous energy-saving change. Payback time for this investment will be a very short period in most areas. In attics that need additional insulation, a layer of batts or blankets can be added over the old insulation. The new insulation *must not* have a vapor barrier. This type of insulation is called "unfaced." Poured insulation can also be used for attics. This type is poured out of packages onto the attic floor and then leveled with a rake or board.

Walls that are in open areas with access to the framing stud wall are easy to insulate. Blankets should be fit snugly against the top portion of the framing material. The insulation should then be neatly stapled to the studs, with staples about 8 inches apart. The blanket should be cut to fit very snugly against the bottom portion of the material. When more than one piece of blanket is used in the same space, the pieces must be neatly cut and fitted tightly together. *It is important that the vapor barrier face the wall that is heated in the winter season.* Insulation should be carefully placed behind electrical boxes, pipes, and ducts by cutting small pieces to the proper size. It is possible to insulate masonry walls by first placing vertical wood strips along the wall and then proceeding. There are types of blanket insulation specially made for use on masonry walls.

Floors that are above cold spaces are insulated by placing blankets or batts between floor joists. The vapor barrier should be up. For holding the insulation in place, wire cut to size to fit snugly across the joists can be used. Also, wire laced back and forth and held by nails provides good support. Insulation should be carefully fitted, with vapor barriers in, along the sills at the ends of the floor structure.

It is not always possible to install your own insulation in a building. Blown insulation can be installed by a contractor in unexposed wall areas and in attic spaces. Care should be taken to assure that the proper *R*-value of blown insulation is provided. Blown insulation and packages of loose-fill insulation must be labeled to show the minimum thickness for a certain *R*-value or the maximum square foot floor area per package to achieve the needed *R*-value.

Duct and Pipe Insulation. Another area of heat loss that results in wasted energy is around heating ducts and pipes in a building. Insulation should be placed around heating ducts and pipes to avoid excessive heat loss. Some types of duct and pipe insulation are shown in Figure 3-8.

Roof Insulation. Insulating material can be placed on roofs to help conserve energy in buildings. Sheets of fiberglass roof insulation can be used to insulate flat or low-slope roofs. These sheets can serve as a base for built-up roofing. In some climatic areas the installation of roof insulation would have a short payback period.

DESIGN TEMPERATURE DIFFERENCE

Another factor to be considered in the study of building structures is called *design temperature difference* (DTD). This is the difference between inside and outside temperatures in degrees Fahrenheit. The outside temperature is considered to be the lowest temperature that is expected to occur several times a year. The inside temperature is the desired temperature (thermostat setting). DTD is important when calculating the heating and cooling loads of a building.

DEGREE-DAYS

A factor used in conjunction with design temperature difference in calculating heating and cooling loads of buildings is called

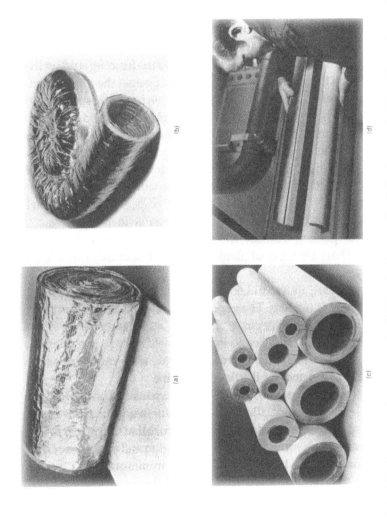

Figure 3-8. (a) Insulation wrap for ductwork; (b) insulation for flexible ductwork; (c) various sizes of pipe insulation; (d) pipe insulation being installed. (*a-c, courtesy of Certain Teed Corp.; d, courtesy of Armstrong Cork Co.*)

degree-days. The degree-day factor is used to determine the average number of degrees that the mean temperature is above or below 65°F. These data are averaged over seasonal periods for consideration in insulating buildings.

Maps that show the zones of the United States for heating and cooling degree-days are shown in Figure 3-9. These zones show the relative amounts of heating or cooling required in certain areas.

PRODUCTS FOR ENERGY CONSERVATION

Several products are now available to be used with building structural systems to help conserve energy. Some of these products are discussed in the following sections.

Solar Window Film
Solar window film can help reduce energy use in buildings. This film is permanently bonded to the inside of the window glass during its installation. It is similar to tinted glass; however, it rejects more sunlight while saving room heat that escapes through windows during the winter season. Solar window film provides a reflective surface to control heat gain in summer and heat loss in winter. The film functions by reflecting sunlight, thus reducing glare inside. Essentially, solar window film can reduce the heat of sunlight passing into a building, reduce glare inside, and reduce heat loss through windows. The installation of solar window film could provide substantial energy savings in buildings. A building with one type of solar window film installed is shown in Figure 3-10.

Air Doors or Curtains
Air doors or curtains are air circulation units mounted above open areas. They provide an air barrier to protect exterior or interior openings from heat transfer. One major use is in sales areas which have large openings to provide easy access to merchandise for customers. Air curtains provide increased energy savings in these areas.

Insulating Glass Windows
A relatively new idea to help conserve energy in window areas is insulating glass. These windows are made with two or more

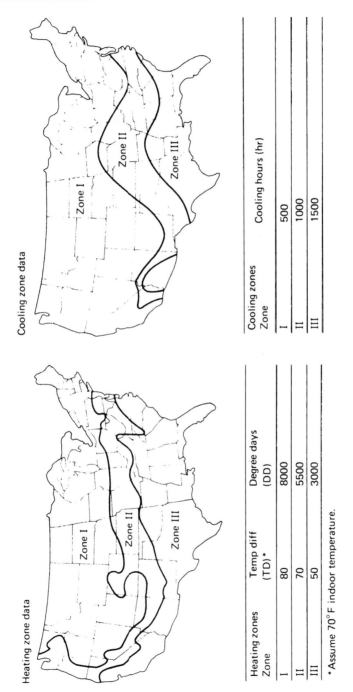

Heating zone data

Heating zones Zone	Temp diff (TD)*	Degree days (DD)
I	80	8000
II	70	5500
III	50	3000

*Assume 70°F indoor temperature.

Cooling zone data

Cooling zones Zone	Cooling hours (hr)
I	500
II	1000
III	1500

Figure 3-9. Maps showing heating and cooling degree-day zones. (*Courtesy of Johns-Manville Sales Corp.*)

Figure 3-10. Building with solar window film. (*Courtesy of Madico Co.*)

panes separated by an air space or thermal break. The thermal break insulates the outer portion of the window from the inner part, thus reducing air infiltration. It is also possible to purchase insulating glass windows with blinds mounted inside the panes. Several examples of insulating glass were shown in illustrations of windows earlier in the chapter.

Window Shades and Blinds

Window shades and blinds, such as those shown in Figure 3-11, can provide energy savings in most climatic areas, as well as beautify window openings. They are designed to lower cooling costs in the summer and heating costs in the winter. Some manufacturers have developed a type of window blind which is designed to operate between two panes of glass. The type shown in Figure 3-11 provides a thermal barrier against heat flow and air currents. They can be operated manually or automatically in response to timers, thermostats, or sun-operated switches.

Figure 3-11. Thermo-shade thermal barriers. (*Courtesy of Solar Energy Components, Inc.*)

Double-Glazed Window Panes

It is possible to reduce heat loss through double-glazed window panes without purchasing complete insulating glass units. Some companies make window panes which can be used to add a second pane on the inside of the existing glass in a building's present windows. The second pane is hermetically sealed to the original glass at the outer part of its frame. This forms an insulating space between the two panes which will reduce heat loss.

Insulated Drapes and Curtains

When drapes and curtains are needed to cover window and door areas, they can be replaced by insulated types. This is another method of reducing heat loss through window and door areas.

Insulated Shutters

Insulated shutters are now made to cover entire window openings. They can reduce heat loss by a substantial amount.

Magnetic Storm Windows

A relatively new development to consider for energy conservation is magnetic storm windows. Some companies now manufacture
window systems which have clear acrylic panels that are attached to
existing windows by magnetic gaskets similar to those used on refrigerator doors. The airtight seal is used to reduce air infiltration and
heat loss.

Door Seals and "Drive-Through" Doors

Many industrial buildings have truck docks which are used
periodically. The large doors used in these areas are usually kept
open for long periods of time. If the area inside is heated or cooled,
this causes a waste of energy. Large amounts of heated or cooled air
escapes through the opening. Door seals are available which can be
used to reduce this energy loss around the doors and compress when
a truck backs up against them. They block the passage of air around
the door and thus conserve energy.

Another method for reducing energy loss through large doors
is the use of "drive-through" doors. These doors are made of soft
insulating material and function similar to Venetian blinds. They are
mounted along the door opening and allow vehicles to drive through
them by opening as a vehicle progresses through.

**ENERGY CONSERVATION CHECKLIST
FOR BUILDING STRUCTURES**

A simplified checklist of things to look for in a building structure
that affect energy consumption is presented in Table 3-1. Some suggestions for changes that can be made to reduce energy use are also
listed.

Table 3-1. Checklist for Building Structures/Saving Money

Items to check	Corrective action
Insulation. Check the amount of insulation in walls, ceilings, and floors. All areas between heated or cooled spaces and outside or unheated or uncooled spaces should have the proper amount of insulation. Areas to check also include storage areas, basements, and attics Also check for water-damaged insulation. Assure that the vapor barrier faces the heated or cooled space and is properly installed.	Have the proper amount of insulation put into the underinsulated areas. Wall insulation can be pumped into wall cavities and between studs. Solid walls can have insulation put on and then be covered. Some minor patching and tacking may also be helpful.
Weather Stripping and caulking. Check around doors windows, piping, and other openings to see if air infiltration can occur.	Use caulking or weather stripping to seal all areas where air infiltration could occur. It is possible to purchase specially designed covers for some open areas.
Doors. Check doors for large air gaps around the outside frame. Self-closing doors should be checked to assure that they close rapidly to prevent large amounts of air infiltration. Consider self-closing doors if they are not used for areas where the door might often be left open to allow air infiltration. Assure that overhead	Attempt to rehang or adjust doors that do not close properly. Adjust the closing mechanism of self-closing doors for rapid closing. Replace defective gaskets or weather stripping on all doors. A device called a "drive-through" door can be installed with overhead doors that have to stay open for long periods of time

Table 3-1. Checklist for Building Structures/Saving Money

Items to check	Corrective action
doors are not left open for excessive periods of time when heating or cooling units are operating.	when heating or cooling units are operating.
Window. Check all windows to see if any are broken or cracked, thus allowing infiltration. Also check to see if they are fitted along their frames properly. Pay particular attention to windows in basements and attics. Consider storm windows, insulating glass, and insulated drapes where they are not used.	Repair or replace broken or cracked window glass. In some cases window areas that are not used can be sealed. Insulating glass separated by air spaces is also available to replace entire window. Storm windows should be insulated wherever possible.
Outside glass areas. Check to see if the building has far too much glass area than is needed for natural lighting and ventilation. Check especially areas such as hallways, corridors, lobbies, and storage areas.	Replace glass areas not needed with insulated wall sections. Insulating glass could be used to replace single-pane glass. These modifications can be very expensive and should be carefully considered before beginning.
Stairwells and other vertical openings. Check stairwells and other vertical openings in the building to see if there are openings that allow air infiltration to heated or cooled areas.	Assure that access doors between areas of different temperatures are always closed and that openings that allow infiltration are sealed. If the space between stairwells and heated or cooled spaces are not closed, doors should definitely be installed.

Chapter 4

Comfort Heating Systems/ Saving Natural Resources

INTRODUCTION

The natural environment of a building is rarely at the right temperature for a person to do work or reside in comfortably for a prolonged length of time. As a general rule, it is either too hot or too cold, depending on the time of day, season, or geographical location of the building. To overcome this type of problem, buildings are usually equipped with an auxiliary heat source or possibly equipped with cool air to maintain the interior environment at a suitable operating level. The comfort heating system of a building is primarily responsible for achieving this function. Efficient design and utilization of heating systems can save natural resources.

THE HEATING SYSTEM CONCEPT

The word system was defined in Chapter 1 as an organization of parts that are needed to form a complete operating unit. The comfort heating system of a building lends itself to this concept. In order to be operational, a system must employ a number of functioning parts. This includes such things as an energy source, a transmission path, control, a load device, and something that will indicate or record the level of heat energy produced. When all of these parts function properly, the system is operational. Figure 4-1 shows a simplified block diagram of a comfort heating system.

The *energy source* of a comfort heating system is essentially responsible for converting energy of one form into something that

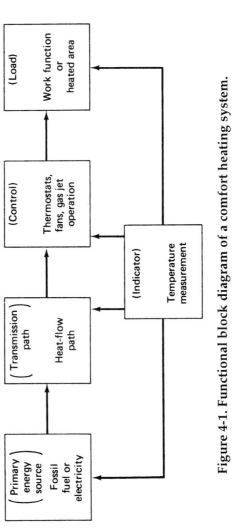

Figure 4-1. Functional block diagram of a comfort heating system.

is entirely new and of a different form. In this regard, the primary source of energy is changed into heat energy. In practice, fossil fuels and electricity are typical primary energy sources.

The process of changing a fossil fuel into heat energy is readily achieved by a chemical reaction called *combustion*. This action takes place when a specified material, mixed with oxygen, results in the production of heat energy. Air is a source of oxygen with coal, natural gas, fuel oil, and liquid petroleum being some of the common fossil fuels.

Electricity may also be used as an energy source for some comfort heating systems. Heat is produced by simply passing an electric current through a resistance element. The resulting heat, and to some extent light energy, is directed away from the element by radiation. Electric heat sources may be used to energize an entire building, as portable sources, or as supplementary heating units. Electric heat is clean because it does not involve burning gases and it is easy to control.

The *transmission path* of a comfort heating system is quite unique when compared with other systems. For example, heat can be transferred through anything that is solid, liquid or gas. If a solid piece of metal is heated at one end, heat will be transferred to the other end by conduction. Liquids and gases are heated primarily by the convection process. This occurs through circulating currents that are developed within the material. Transmission may also occur through radiation. This takes place when rays or waves are emitted from a heated surface and travel through space.

Typical transmission paths for a comfort heating system are small-diameter pipes for steam or hot water, metal ducts or large-diameter pipes for airflow, or radiation paths through the air. The transmission path is directly dependent on the size and design of the heating system.

The *control* function of a comfort heating system is primarily responsible for altering the flow of heat anywhere between the primary energy source and the load device. Circulating fans, for example, are often used to alter the flow of warm air through a system. Simply turning a fan on or off has a pronounced effect on the volume of airflow. The gas jets of a forced-air furnace are also turned on and off according to actual temperature and the set-point adjustment of a thermostat. In comfort heating, control may be achieved manually or automatically, depending on the design of the system.

The *load* of a comfort heating system represents the primary work function or area that is to be maintained at a specific temperature. Since the nature of heat is to flow away from a body of higher temperature and toward a body or substance of a lower temperature, the load of a heating system is always directly dependent on the energy source. The heat demand of the system load may be variable, steady, or cyclic, depending on the building structure.

The *indicator* of a comfort heating system is primarily responsible for measurement operations. In some small building systems and in portable unit heaters, the indicator may be omitted entirely. In larger systems there may be several indicators located at key test points. Indicators, such as thermometers, are usually considered as a working part of the thermostat which senses temperature. System control is normally triggered by changes in temperature sensed by the thermostat. Indicators may also be permanently attached to the system for continuous display of information or they may be installed at key test points for maintenance tests.

TYPES OF HEATING SYSTEMS

The type of heating system used in a building today is dependent primarily on the structure of the building and its intended work function. In residential buildings, schools, churches, commercial facilities, and most industrial buildings, a *central heating system* is more common. A prime location for this system is in a low-level room or near the center of the structure, where distribution is convenient to all parts of the building. Central heating systems are energized by gas, coal, fuel oil, or electricity. Figure 4-2 shows a representative central heating unit that employs natural gas.

Many industrial heating installations do not require ducts or a central distribution path to deliver heat to specific rooms or areas of a large building. Smaller, more efficient self-contained heating systems called *unit heaters* are used to achieve this function. This type of system has essentially the same functional parts as does a larger central heating system but on a smaller scale. Many installations of this type

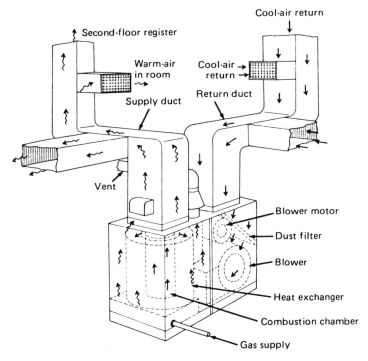

Figure 4-2. Natural gas forced-air comfort heating system. (Courtesy of American Gas Association Inc.)

employ ceiling-mounted units that direct their heat toward a specific work area or room. Unit heaters derive primary energy from gas, fuel oil, steam, or electricity. These heaters range in size from small portable units to rigidly installed assemblies that are attached to the building. Figure 4-3 shows a representative vertical delivery electric unit heater assembly.

FOSSIL-FUEL HEATING SYSTEMS

Fossil fuel is one of the basic ingredients of the energy source of a large majority of the building comfort heating systems in operation today. In this respect coal, natural gas, liquid petroleum, fuel oil, or a combination of these fuels is burned to produce heat. Operation of the system regardless of the fuel used is basically the same, with only a few

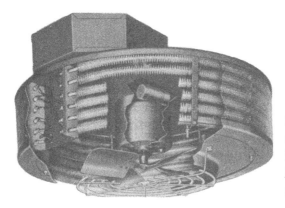

Figure 4-3. Vertical-delivery electric unit heater. (*Courtesy of Modine Manufacturing Co.***)**

minor modifications. The essential components of this type of heating systems are the combustion chamber, heat exchanger, venting components, control unit, air blower, fuel ignition, and duct networks.

FORCED-AIR GAS FURNACES

Forced-air furnaces have been used for a number of years as comfort heating systems for commercial, industrial, and residential buildings. The primary difference in a specific unit selected for an application is based upon its heated airflow capacity. The size and Btu rating of the burners, heat-exchanger design, and blower capacity are the primary differences in heating system size.

Figure 4-4 shows a cutaway view of a representative forced air gas furnace. Note in this example the location of the burners, heat exchanger, blower, and its electronic ignition unit. Each of these components is discussed next in some detail together with the general operation of the system.

Figure 4-4. Forced-air gas furnace. (*Courtesy of Lennox Industries, Inc.***)**

Natural Gas Fuel

Natural gas is the lightest fuel of all petroleum products. Most theorists believe that natural gas is derived from decayed plant and animal remains that were buried in prehistoric times.

Natural gas and petroleum are the result of hydrogen (H) and carbon (C) that existed in prehistoric plants and animals. The term *hydrocarbon is* often used to describe these products. Natural gas specifically contains 55 to 98% methane (CH_4), 0.1 to 14% ethane (C_3H_8), and 0.5% carbon dioxide (CO_2). When used as a fuel, approximately 15 cubic feet of air is needed with each cubic foot of gas to produce combustion. Because of the low boiling point of the methane and ethane components, natural gas continues to remain a gas even under high pressure and temperature. The heating capability or Btu rating of natural gas varies quite readily between different areas and localities. The local gas company will generally supply typical Btu ratings of the actual gas supplied to a specific system.

Natural Gas Burners

The gas burner of a furnace or heating system is primarily responsible for the mixing of air and gas in a proper ratio to ensure combustion. The 15:1 air/gas ratio is quite common in most geographical areas of the United States.

The design of a gas burner is primarily responsible for the mixing of gas and air in proper proportions. The first part of this design usually includes a slotted port which is used to admit air into the gas stream. A burner mixing tube then combines the two as they pass through an area of reduced diameter called a *venturi.* This action provides a constant mixture and an even flow of gas and air into the burner head. The burner head then releases the mixture into the combustion chamber. Burner head types include drilled ports, slotted ports, ribbons, and an inshot design as shown by Figure 4-5. The continuous slotted port burner of Figure 4-6 is widely used in small- and medium-size gas-burning heating systems today.

The amount of air admitted to the burner assembly is normally controlled by a manual shutter on the front of the burner as in Figure 4-6. Adjustment of this shutter either enlarges or reduces the opening, according to its position. The color of the resulting burner flame, as a rule, is a good indication of the resulting fuel/air mixture ratio. A

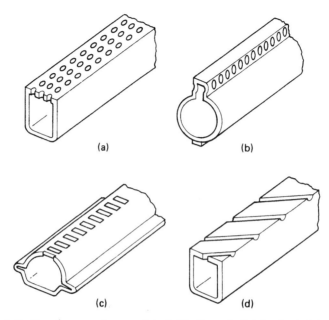

Figure 4-5. Gas burner types: (a) drilled port; (b) inshot; (c) slotted post; (d) ribbon.

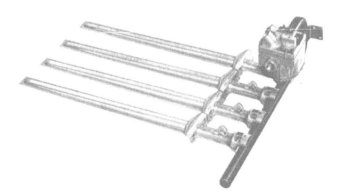

Figure 4-6. Continuous slotted port gas burner. (*Courtesy of Carrier Corporation.*)

yellow flame, for example, denotes insufficient primary air, whereas a sharp blue flame shows a proper mixture ratio. Burner efficiency is directly dependent upon the fuel/air mixture ratio.

Forced-draft Burners

To be assured of a more consistent airflow for the combustion process, larger gas furnaces are now being equipped with forced-draft burners. The combustion area of this unit is essentially sealed from the atmosphere and air is forced into it under pressure.

The amount of air supplied to the chamber is controlled by an adjustment of the flue outlet. With a fixed amount of air and gas applied to the burner assembly, combustion level and mixture are easier to control.

Forced-draft burners are frequently found today in roof-mounted unit heaters and in larger industrial furnaces. Figure 4-7 shows a sketch of a forced-air burner assembly.

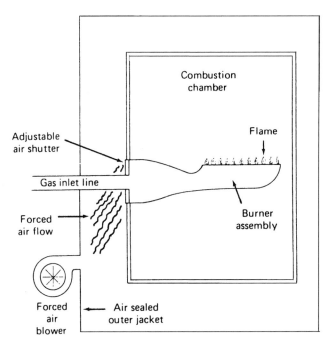

Figure 4-7. Forced-draft burner assembly.

Pilot Burner and Pilot Safety Device

The pilot burner and pilot safety device of a gas-fired furnace play an extremely important role in normal system operation. The pilot burner is, first, responsible for ignition of the gas/air mixture that is emitted from the main burner. Second, the pilot safety device is used to detect the loss of pilot burner operation. As a result of this action, gas will not be supplied to this main burner when it cannot be properly ignited. This feature provides a fail-safe measure that prohibits unburned gas from accumulating in the combustion chamber.

The pilot safety device is essentially designed to respond to heat energy. Typically, a thermocouple is positioned in the burning pilot flame as in Figure 4-8. This device simply changes heat energy into a small electrical voltage. The output voltage of the thermo-couple causes current to flow in the solenoid coil of the main gas value. If for some unforeseen reason, the temperature of the thermocouple drops below 3000°F, no voltage will be developed. As a result of this, the solenoid valve will close down. Through this procedure, the pilot assembly will prevent a dangerous accumulation of unburned gas. More than half of all gas-fired heating system failure can be attributed to pilot burner problems.

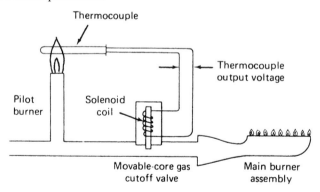

Figure 4-8. Pilot safety control of a gas burner.

Electronic Flame Ignition

Electronic flame ignition is a rather new development in gas heating systems. This type of unit develops a number of high-voltage discharging sparks for ignition of the pilot burner. Figure 4-9 shows

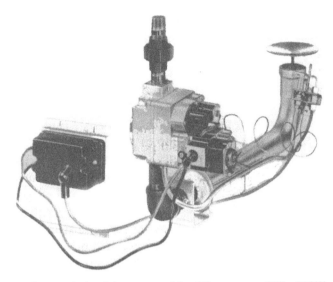

Figure 4-9. Electronic ignition assembly. (Courtesy of The Williamson Co.)

the pilot ignition, main burner unit, and the electronic spark assembly used in a gas furnace.

To see how pilot ignition is produced electronically, let us consider what occurs in a typical operating sequence. Assume first that the thermostat responds by closing contact points, indicating the need for more heat. This action immediately turns on the electronic-pulse generator circuit and the pilot burner gas valve. Low-voltage pulses, approximately 100 per minute, are immediately applied to a transformer, where they are stepped up to a value of 10,000 volts. If the pilot does not ignite by the sparks within approximately 30 seconds, the entire unit shuts off as a fail-safe measure. Normal ignition of the pilot causes it to continue operation. After a few seconds of pilot operation, enough heat is developed to actuate a thermal warp switch. This operation is then used to turn on the main gas burner solenoid value. This same action, in turn, shuts off the pulse generator circuit and pilot burner. The main burner continues to operate until the thermostat setting is satisfied. It then turns off and waits in a ready state for the next command signal from the thermostat.

Electronic pilot ignition is finding widespread usage today in new furnace installations. Through this type of ignition, it is possible to have pilot burner safety protection, as no gas is consumed when the furnace is not in operation.

Heat Exchangers

The heat exchanger of a gas-fired system is primarily responsible for the transfer of combustion heat to the heating medium. In practice, there are two paths through which heat energy flows. The main burner is placed inside one flow path, designed to circulate heated gas after combustion occurs. This gas simply passes through the inside of the exchanger and is ultimately expelled into the flue. Air or the heating medium takes an alternate route as it passes around the outside of the exchanger. Heat transfer in this case takes place from the inside to the outside of the exchanger. Figure 4-10 shows the heat exchanger of a low-capacity gas furnace.

A cutaway view of the furnace in Figure 4-11 shows an installed heat exchanger. The sculptured curve of this unit is designed to provide free expansion and contraction without a cracking noise. Combustion gases passing through the smooth curves produces efficient heat transfer without noise. The sculptured curved area also provides some restriction in the combustion chamber so that an undue amount of gas will not escape into the flue without heat transfer. Through this type of design a transfer of heat of rather high efficiency will take place.

A sectionalized heat exchanger is shown in Figure 4-12. In this type of unit, main burners are placed in openings at the bottom of the assembly. After passing through the internal part of the chamber, gas fumes enter the flue through top openings. System air then circulates over the outside surfaces of the exchanger, where it gains heat before being returned to the duct network. The unit shown is ceramic-coated to improve heat transfer, reduce rust, and prevent metal expansion and contraction noise.

In practice, heat exchangers provide trouble-free operation for long periods of time. With proper installation of the burners and correct adjustment of the air/gas ratio, the inside of the chamber will remain nearly free of carbon deposits. However, a small hole or welding break in the heat exchanger wall can cause a very serious condition to occur. Carbon monoxide gas will be admitted into the heated air-

Figure 4-10. Heat exchanger. (Courtesy of Lennox Industries, Inc.)

Figure 4-11. Installed heat exchanger. (Courtesy of Lennox Industries, Inc.)

Figure 4-12. Sectionalized heat exchanger. (Courtesy of Carrier Corporation.)

stream through the break or hole. As a rule, problems of this type can be detected by careful inspection of the combustion chamber with a light source.

Draft Diverter

After fuel is burned in the combustion chamber and passes through the heat exchanger, the remaining gases must be vented into the outside air. Upon leaving the heat exchanger, these gases first enter a draft diverter assembly. At this point, gases are mixed with a quantity of air that is essentially equivalent to the amount needed for combustion. Figure 4-13 shows the draft diverter and air relief opening of a forced-air gas furnace.

The draft diverter is designed to neutralize any excessive updrafts or downdrafts that might occur through the flue of a heating system. This means that a continuous source of air will be provided regardless of the draft status of the flue. Abnormal drafts adversely

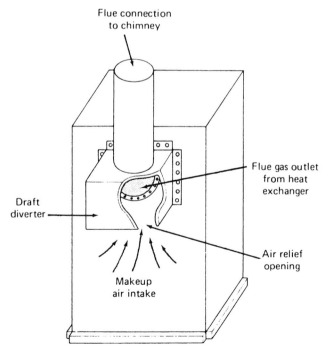

Figure 4-13. Draft diverter and air-relief opening of a gas furnace.

affect the efficiency of a furnace and cause excessive generation or harmful carbon monoxide gas. This action may eventually lead to ignition problems, pilot light extinguishing, and unnecessary shutdowns. Through proper installation of the draft diverter, continuous operation of the unit is assured by providing consistent air/gas mixtures.

In practice, the draft diverter of a gas furnace should not be physically changed or altered after the system has been installed. Changes in this part of the system adversely affect operation and system safety.

Venting

The venting section of a gas-fired heating system is primarily responsible for transporting spent gases away from the unit and into the outside atmosphere. In achieving this function a vent system must simply provide a path of minimum resistance for flue gases and serve as a source of dilution air for the draft diverter. Design of the vent pipe must take into account such things as total combustion chamber area and pipe height.

The term *draft* is commonly used to describe a force applied to the movement of air as it passes through a heating system. The force involved here is based upon a difference in pressure that is developed by the weight of the earth's atmosphere at different locations.

For a draft to be produced, pressure developed by the air surrounding the heating unit and vent must be greater than that produced by burning gases within the unit. This is achieved primarily by a temperature difference that occurs between inside and outside air. The height to which this temperature difference occurs is also quite important. Vent pipe height determines this condition. A suitable draft is usually developed in gas furnaces by attaching the flue outlet of the heat exchanger to a vent pipe or stack that extends above the roof of the building.

Heating System Control

The control function of a heating system is primarily responsible for heating a building to a comfortable level and maintaining the temperature at a predetermined value. In practice, this system function is usually achieved through the use of low-voltage electricity. Typical control functions apply to the blower fan, main gas valve, and the temperature-limit control. A thermostat is used to control the main gas valve; other valves are controlled by thermal warp switch action.

Blower fan control responds to temperature changes that occur in normal heating system operation. A temperature-sensing element, such as a warp switch, is normally located inside the heat exchanger. When this element senses a rise in temperature, it closes a set of contact points. This action, in turn, completes an electrical circuit to the blower motor which causes it to turn on. When the heat-exchanger temperature drops to a certain value, the warp switch opens its contacts and turns off the blower motor. Typical fan control is set for an operating range of 150°F on and 100°F off. Figure 4-14 shows a direct-drive blower fan and the air circulation path that occurs when it is in operation.

Figure 4-14. Gas furnace blower motor assembly. (Courtesy of Lennox Industries, Inc.)

The main gas valve is controlled electrically by action of the thermostat and pilot safety valve. When heat is needed to meet the demands of the thermostat, the main gas burner valve is actuated if the pilot safety valve is operational. Normally, when the main valve is actuated it admits gas to the combustion chamber. In practice, a number of control functions are achieved by a single self-contained assembly. Gas pressure regulation, pilot safety, main gas cock adjustment, pilot gas cock adjustment, and the main solenoid valve are all an integral part of this unit. Figure 4-15 shows an enlarged view of a gas valve and its installation in an operating unit.

The temperature limit control is a heat-actuated switch that is placed inside of or attached to the heat exchanger. This switch is then connected to the primary ac line voltage feeding the entire furnace control assembly. Should the temperature inside the heat exchanger reach or exceed a predetermined upper limit value, the source voltage will be turned off. Through this type of control the furnace is limited to a temperature operating range that will not cause damage to the combustion chamber. A limiting range of 200°F is commonly used in low-capacity gas-fired furnaces today.

Figure 4-15. Main gas valve installation. (Courtesy of The Williamson Co.)

Thermostat

The thermostat of a heating system is located in a prime sampling area of the building to sense system temperature. A set-point value is adjusted by turning a dial. When the temperature drops below the set-point value by a few degrees, the thermostat senses this change and turns on a set of contact points. This action calls for more heat by turning on the main gas burner and eventually the blower fan. When the set-point temperature value is realized, the burner solenoid valve is turned off electrically. The thermostat generally achieves control alteration of a warp switch that is altered by the manual set-point adjustment. Figure 4-16 shows a representative thermostat and its electrical control circuit. Advanced designs of thermostat technology can be found in Chapter 11, Energy Management.

Heat Distribution Network

The delivery of heat from a gas-fired system generally involves a network of distribution ducts. When a heat exchanger develops heat from the combustion chamber, it is circulated through the system by a blower fan. Cool air is initially pulled into the system through ducts placed at strategic floor-level locations. This cool air first passes through a filter, where dust and foreign particles are removed. It then circulates around the heat exchanger, where it is warmed and returned to each room through the network of supply ducts.

The heat distribution network of a gas-fired furnace essentially represents a supply of warm air from its output with cool air returned to its input when the blower fan is in operation. As a result of this action, work is accomplished by the furnace. The composite function of heat production and transfer represents the loading of the system. The heating capacity of the system is based upon its ability to do work in altering the load. Figure 4-17a shows a partial view of a gas-fired forced-air comfort heating system for a small building.

HIGH-EFFICIENCY GAS FURNACES

As energy costs for natural gas have risen, the need for improving the efficiency of the standard gas furnace has risen. The standard measure of efficiency that compares the amount of heat a furnace

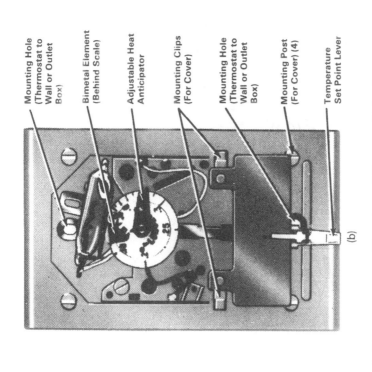

Mounting Hole (Thermostat to Wall or Outlet Box)

Bimetal Element (Behind Scale)

Adjustable Heat Anticipator

Mounting Clips (For Cover)

Mounting Hole (Thermostat to Wall or Outlet Box)

Mounting Post (For Cover) (4)

Temperature Set Point Lever

(b)

Honeywell

• 60 • 70 • 80 • 90 •

(a)

Figure 4-16. (a) Thermostat; (b) components. (*Courtesy of Honeywell, Inc.*)

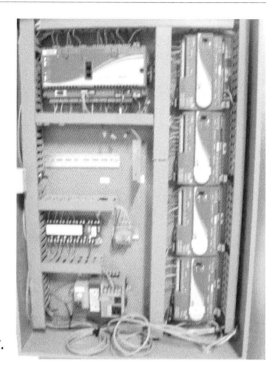

Figure 4-17.

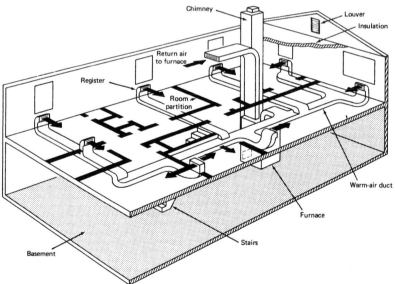

Figure 4-17a. Forced-air gas-fired heat distribution network.

delivers to the amount of fuel supplied is the Annual Fuel Utilization Efficiency or AFUE rating. In 1992, the US Department of Energy mandated that all furnaces sold must have a minimum AFUE rating of 78%. Furnaces with an AFUE rating of 78% to 82% are generally titled 'mid-efficiency' furnaces. Those furnaces that exceed the 88% rating are considered 'high-efficiency' furnaces. This increased efficiency is due to a handful of changes in technology. The first change is the elimination of a pilot light that burns fuel even when there is no demand to an electronic ignition. The second is to draw combustion air from outside the structure, thereby not expelling any air that has already been heated. The most significant change in technology is the ability to extract additional heat from the natural gas so efficiently that the flue gasses condense, or turn vapor into water. It is this ability to condense flue gasses that gives high-efficiency gas furnaces the nickname condensing furnace. These high-efficiency furnaces (see Figure 4-17b) run the spent fuel gasses that would normally be expelled through a chimney of flue through a second heat exchanger. This second heat exchanger extracts additional heat from these gasses

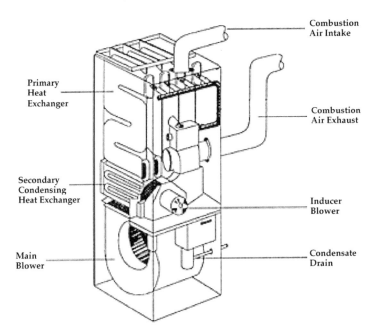

Figure 4-17b. High efficiency condensing furnace.

that would normally be sent up the flue. This process involving the second extraction leaves flue gasses at such a low temperature that water is condensed and must be drained or pumped away from the system. Because of the low temperature of the flue gasses, venting these gases do not require masonry or triple-wall flues, the gasses can be carried by plastic pipe, thereby making the high-efficiency furnace less expensive to install.

FUEL OIL-BURNING SYSTEMS

Fuel oil is a very common form of fossil fuel that can be effectively burned to produce heat. Comfort heating systems that employ this type of fuel are found widely in commercial, industrial, and residential buildings today. The use of fuel oil is primarily the result of convenience and the cleanliness of oil when compared with coal. In the northeastern section of the United States, fuel oil is a major source of energy for many comfort heating systems.

Fuel oil is a by-product of petroleum refining. It is graded according to the amount of distillation performed during its production. Grades range from 1 to 6 with the number 3 omitted. Lower-grade numbers contain fewer impurities and are more expensive. Number 1 and 2 oil is best suited for comfort heating system applications. No. 1 fuel oil produces 133,000 to 137,000 Btu/gal, with No. 2 developing approximately 136,000 to 142,000 Btu/gal. Higher-numbered oil grades are thicker, produce higher Btu/gal values, and are used primarily in industrial heating applications.

Heat produced by natural gas or fuel oil is achieved by essentially the same basic type of system components. This includes such things as a burner, fuel igniter assembly, combustion chamber, heat exchanger, thermostatic control, venting, a blower fan, and a duct network. Figure 4-18 shows a forced-air fuel oil-fired comfort heating furnace with its outside jacket removed. Central systems of this type range in heating capacity from 85,000 to 200,000 Btu per hour. In addition to its use in the central heating type of system, fuel oil is used to energize unit heaters in industrial and commercial buildings. Figure 4-19 shows an assembled oil-fired unit heater.

Figure 4-18. Forced-air oil-fired furnace. (*Courtesy of The Williamson Co.*)

Figure 4-19. Oil-fired unit heater. (*Courtesy of Modine Manufacturing Co.*)

Fuel Oil Burners

The burner assembly of a fuel oil furnace represents the primary difference between gas- and oil-fired heating systems.

The basic reason behind this difference is in the physical state of the fuel being used. Fuel oil in its liquid state simply does not burn very effectively. For burning to take place, it must either be heated to produce a vapor or mixed with air. Vaporizing is rarely used in comfort heating systems because it is somewhat difficult and expensive to achieve. Atomizing, which is a fuel/air mixing procedure, is used in nearly all comfort heating system installations today.

A fuel oil burner assembly is used primarily for combustion. The mechanical function of this unit is responsible for breaking oil into a fog-like mist. This can be achieved by (1) forcing air and oil under pressure through a nozzle, (2) forcing oil out the end of small tubes attached to a rotating shaft, or (3) passing oil over the lip of a rapidly rotating cup. In practice these methods must either employ an electric motor to produce shaft rotation or develop air pressure for the mixing process. A large number of these burners are small compact assemblies that can be readily attached to the combustion chamber. Figure 4-20 shows a representative low-pressure oil burner assembly unit.

Pressurized burners are made up of an electric motor, a blower, a pump, and a fuel unit. The blower is attached to the motor shaft and develops air for mixing with the oil. A fuel pump is used to develop pressure according to the design of the system. Low-pressure units respond to a range of 1 to 15 psi, whereas high-pressure systems may reach levels of 100 psi. The fuel unit contains a pressure-regulat-

Figure 4-20. Fuel oil burner assembly. (*Courtesy of The Carlin Company*)

ing valve and a filter for straining foreign particles from the fuel.

In an operational sequence, when heat is called for by the thermostat, a solenoid-controlled fuel-level valve opens to admit fuel into the line from the storage tank. It first passes through a filter strainer where any existing solid particles are removed from the fuel before it is applied to the pump. Outflow of the pump must be at a selected pressure level according to the design of the system. An in-line pressure-regulating valve follows the pump and maintains the pressure at a desired operating level. Fuel then proceeds under pressure to the fuel nozzle.

This nozzle simply changes pressurized fuel into tiny droplets which are ejected and mixed with air from the blower. A 10,000-volt spark developed by a high-voltage transformer is used to ignite the atomized fuel at this point. The burning process starts and is continuous as long as fuel is being supplied. When the heat demand of the thermostat setting is met, fuel flow ceases and the burner shuts down. The unit then remains in a ready state waiting for the next heat demand.

A second basic type of oil burner assembly employs the rotary principle. *Rotary oil burners* are quite compact and usually have a rather limited number of moving parts. Oil in this case is applied to a hollow metal shaft that turns during normal motor operation. The same rotary action of the motor shaft is also used to turn the blower of the air source. The unique feature of this unit is that it does not necessitate pressurized oil to function. The rotating shaft may employ small tubes or a cup structure for oil emission. Figure 4-21 shows a cross-sectional view of a rotary oil burner with an atomizing cup.

The heat exchanger, blower, duct network, and draft assembly are quite similar to those of a comparable heating system. Figure 4-22 shows a partial cutaway view of an oil-fired unit heater. Note particularly the location of the oil burner, burner nozzle, blower fan, and heat exchanger.

An assembled heat exchanger of an oil-fired unit heater is shown in Figure 4-23. Note the location of the flue collector, service inspection door, and combustion chamber clean-out ports. For safety reasons, inspection doors and clean-out ports of this type are not found on comparable gas-fired heat exchangers.

Heat transfer of a fuel oil-fired unit heater is from the inside of the exchanger to the outside. Cool air is pulled into the assem-

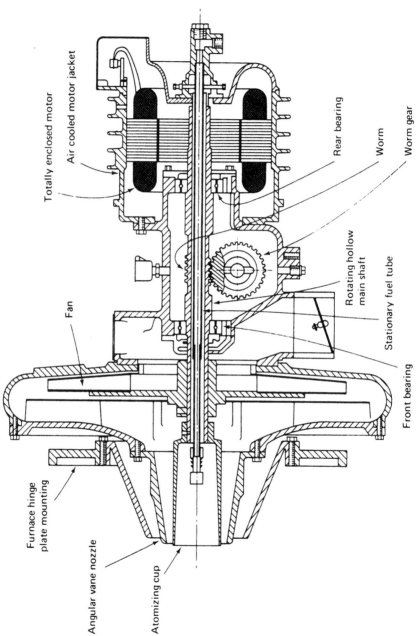

Figure 4-21. Cross-sectional view of a rotary oil burner. (*Courtesy of Ray Burner Co.*)

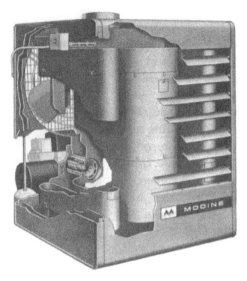

Figure 4-22. Cutaway view of an oil-fired unit heater. (*Courtesy of Modine Manufacturing Co.*)

bled unit from the back side and circulated over the outside of the

exchanger. It exits from the unit through louvers mounted on the front of the assembly. Figure 4-24 shows the location of the blower fan and burner assembly on the back side of the unit.

COAL-BURNING
HEATING SYSTEMS

Coal is a solid type of fossil fuel that has been used for a number of years as an energy source in heating systems. Systems of this type produce heat through the burning of fuel after it has been placed on a metal grate inside the

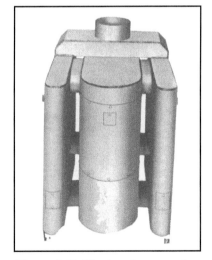

Figure 4-23. Heat exchanger of an oil-fired unit heater. (*Courtesy of Modine Manufacturing Co.*)

Figure 4-24. Rear view of an oil-fired unit heater. (*Courtesy of Modine Manufacturing Co.*)

combustion chamber. The transfer of heat energy in this case is from the inside of the exchanger to the outside. As a general rule, coal-fired furnaces are much larger than other heating systems, have less control, and are somewhat inconvenient to operate. Utilization of these units in commercial and residential buildings has dropped quite significantly in recent years, although industrial applications are continuing at about the same rate.

The fundamental parts of a coal-fired heating system are very similar in many respects to those of other fossil fuel-burning units. Fuel is loaded into the combustion chamber either manually or automatically. Air is mixed with the fuel to produce combustion and is expelled through the draft diverter into an outside stack or flue. Heat developed through the combustion process is then transferred to circulating air that passes around the outside of the exchanger. The resulting output of the heat exchanger eventually passes through a duct network, where it is distributed throughout the facility. Heated air may be moved by a motor-driven blower or distributed through ducts by natural circulation. Figure 4-25 shows the essential parts of a coal-fired heating system.

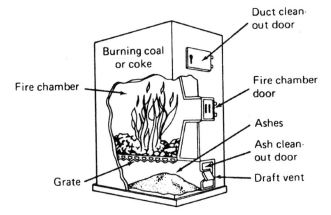

Figure 4-25. Hand-fired coat furnace.

Coal-fired furnaces have a number of features that distinguish them from other heating systems. For example, the fuel is loaded into the system either by hand or automatically by a stoker. Once the burning process is started, it should be continuous. Coal must be supplied periodically to prevent the fire from going out. A burnout necessitates restarting the fire with paper and wood to reach a flashover point that will produce ignition of the new supply of coal. Unburned particles or residue drop into an ash pit under the grate. This must be removed on a regular basis to make the process continuous.

The burning process and heat regulation of a coal-fired unit are also unique compared with other heating systems. Burning is controlled by the loading of new coal and the regulation of air admitted to the combustion chamber through a damper assembly. Closing the damper reduces the flow of air and slows down the burning process. More coal is burned in a given unit of time when the damper is open. Many systems employ a thermostatically controlled damper motor that regulates airflow into the chamber according to the heat demand of the system.

Coal-fired heating systems are not used as much today as they were in the earlier part of the century. Most of this can be attributed to the inconvenience of operation, inefficiency, fuel storage problems, air pollution, and fuel costs. The availability of coal, wood, and other solid fuels compared with gas and oil may cause a rather significant change in the future of solid-fuel-fired heating systems. A number of manufacturing concerns are now developing multi-fuel heating systems that will burn gas, fuel oil, or solid fuel according to its availability. Systems of this type provide a backup or alternative in the case of a primary fuel shortage. Figure 4-26 shows a sketch of a representative multi-fuel heating system.

ELECTRIC HEATING SYSTEMS

Electrically energized comfort heating systems are widely used today to produce heat for commercial, industrial, and residential buildings. This type of energy is readily available at nearly any building site and has a number of advantages over the fuel-burning methods of producing heat. Ecologically, any fuel needed to produce electricity is burned or consumed at a power plant that is usually located some distance away from the building where it is being used. Through this method of heating there is less pollution than there

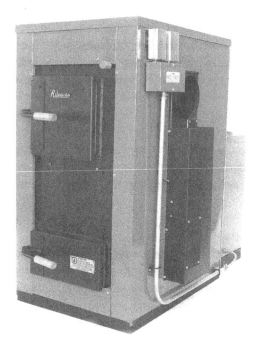

Figure 4-26. Multifuel forced-air heating system. (*Courtesy of Riteway Manufacturing Co.***)**

would be if fuel is burned at each building. Electric heat is also clean to use and easy to control and has a high degree of operational efficiency.

Electric heating is important today because of its high level of efficiency. Theoretically, when electrical energy is applied to a system, virtually all of it is transformed into heat energy. Essentially, this means that, when a specified amount of electricity is applied, it produces an equivalent Btu output. One thousand watts or 1 kW of electricity, when converted to heat, produces 3412 Btu of heat energy.

Heating can be achieved in a variety of ways through the use of electricity. In comfort heating systems, we need only be concerned with such things as resistance heating and the heat-pump principle. These two divisions of equipment both follow the basic systems concept discussed previously. This means that they contain an energy source, transmission path, control load device, and the possibility of one or more optional indicators. The primary difference in the two electrical systems is in the production of heat energy. Resistance heating is accomplished by passing an electrical current through wires or conductors to the load device. By comparison, the heat pump operates by circulating a gas or liquid through pipes that connect an inside coil to an outside coil. Electricity is needed in both cases as an energy source to make the respective systems operational.

Resistance Heating

When an electric current flows through a conductive material, it encounters a type of opposition called *resistance*. In most circuits this opposition is unavoidable to some extent because of the material of the conductor, its length, cross-sectional area, and temperature. The conductor wires of a heating system are purposely kept low in resistance to minimize heat production between the source and the load device. Heavy-gauge insulated copper wire is used for this part of the system.

The load device of a resistance heating system is primarily responsible for the generation of heat energy. The amount of heat developed by the load is based upon the value of current that passes through the resistive element. Element resistance is purposely designed to be quite high when compared with the connecting wires of the system. An alloy of nickel and chromium called Nichrome is commonly used for the heating elements. Construction includes spring-like coils, ribbon elements, and tubular rod elements to improve the heat transfer. Figure 4-27 shows the finned tubular rod element of an electric unit heater. A cutaway view of the unit heater in Figure 4-28 shows the location of the element in front of a circulating fan.

Resistive elements may be placed under windows or at strategic locations throughout the building. In this type of installation the elements are enclosed in a housing that provides electrical safety and efficient use of the available heat. A cutaway view of a baseboard convector section is shown in Figure 4-29. The heating element of this unit is of the finned rod type. Air entering at the bottom of the unit circulates around the fins to gain heat and exits at the top. Different configurations of the unit may be selected according to the method of circulation desired, unit length, and heat-density production.

Resistive elements are also used as a heat source in forced-air central heating systems. In this application, the element is mounted directly in the main airstream of the system. The number of elements selected for a particular installation is based upon the desired heat

Figure 4-27. Heating element of an electric unit heater. (*Courtesy of Modine Manufacturing Co.***)**

Figure 4-28. Cutaway view of an electric unit heater. (*Courtesy of Modine Manufacturing Co.*)

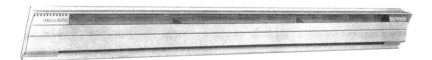

Figure 4-29. Baseboard convector heater. (*Courtesy of Federal Pacific Electric Co.*)

output production. Individual elements are generally positioned in a staggered configuration to provide uniform heat transfer and to eliminate hot spots. Figure 4-30 shows the resistive element of a heat pump. The element of this unit has spring-coil construction supported by ceramic insulators. Units of this type provide an auxiliary source of heat when the outside temperature becomes quite cold. Air circulating around the element is warmed and forced into the duct network for distribution throughout the building.

Figure 4-30. Resistance heating element. (*Courtesy of The Williamson Co.***)**

Heat Pump

A *heat pump* is defined as a reversible air-conditioning system that transfers heat either into or away from an area that is being conditioned. When the outside temperature is warm, it takes indoor heat and moves it outside as an air-conditioning unit. Operation during cold weather causes it to take outdoor heat and move it indoors as a heating unit. Heating can be performed even during cold temperatures because there is always a certain amount of heat in the outside air. At 0°F (–22°C), for example, the air will have approximately 89% of the heat that it has at 100°F (38°C). Even at sub-zero temperatures, it is possible to develop some heat from the outside air. However, it is more difficult to develop heat when the temperature drops below 20°F (–6°C). For installations that encounter temperatures colder than this, heat pumps are equipped with resistance heating coils to supplement the system.

A heat pump, like an air conditioner, consists of a compressor, an outdoor coil, an expansion device, and an indoor coil. The compressor is responsible for pumping a refrigerant between the indoor and outdoor coils. The refrigerant is alternately changed between a liquid and a gas, depending upon its location in the system. Electric fans or blowers are used to force air across the respective coils and to circulate cool or warm air throughout the building.

A majority of the heat pumps in operation today consist of indoor and outdoor units that are connected together by insulated pipes or tubes. Figure 4-31 shows a typical heat pump of this type that is used in a central distribution system. The indoor unit houses the supplemental electric heat elements, blower and motor assembly, electronic air cleaner, humidifier, control panel, and indoor coil. The outdoor unit is covered with a heavy-gauge steel cabinet that encloses the outdoor coil, blower fan assembly, compressor, expansion device, and cycle-reversing valve. Both units are designed for maximum performance, high operational efficiency, and low electrical power consumption.

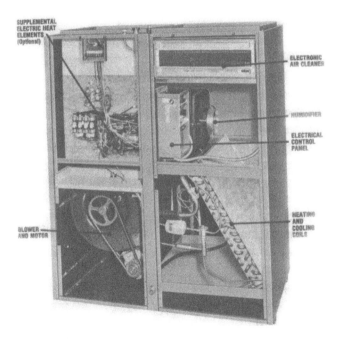

Figure 4-31. Heat-pump system. (*Courtesy of The Williamson Co.*)

The Heat Pump

If a unit air conditioner were turned around in a window during its operational cycle, it would be extracting heat from the outside air and pumping it into the inside. This condition, which is the operational basis of the heat pump, is often called the *reverse-flow air-conditioner principle*. The heat pump is essentially "turned around" from its cooling cycle by a special valve that reverses the flow of refrigerant through the system. When the heating cycle occurs, the indoor coil, outdoor coil, and fans are reversed. The outdoor coil is now responsible for extracting heat from the outside air and passes it along it the indoor coil, where it is released into the duct network for distribution.

During the heating cycle, any refrigerant that is circulating in the outside coil is changed into a low-temperature gas. It is purposely made to be substantially colder than the outside air. Since heat energy always moves from hot to cold, there is a transfer of heat from the outside air to the cold refrigerant. In a sense, we can say that the heat of the cold outside air is absorbed by the much colder refrigerant gas.

The compressor of the system is responsible for squeezing together the heat-laden gas that has passed through the outside coil. This action is designed to cause an increase in the pressure of the gas that is pumped to the indoor coil. As air is blown over the indoor coil, the high-pressure gas gives up its heat to the air. Warm air is then circulated through the duct network to the respective rooms of the system.

When the refrigerant gas of the indoor coil gives up its heat, it cools and condenses into a liquid. It is then pumped back to the outside coil by compressor action. Once again it is changed into a cool gaseous state and is applied to the outside coil to repeat the cycle. If the outside temperature drops too low, the refrigerant may not be able to collect enough heat to satisfy the system. When this occurs, electric-resistance heaters are energized to supplement the heating process. The place where electric heat is supplied to the system is called the *balance point*.

Figure 4-32 shows an illustration of heat-pump operation during its heating cycle. At (1), the heat is absorbed from the cold outside air by the pressurized low-temperature refrigerant circulating through the outside coil. At (2), the refrigerant is applied to the compressor and compressed into a high-temperature, high-pressure gas. At (3),

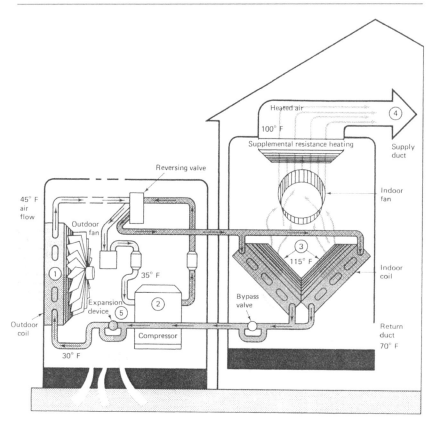

Figure 4-32. Heating-cycle operation. (*Courtesy of Public Service Electric and Gas Company.*)

the heated gas is transferred to the indoor coil and released as heat. At (4), warm air is circulated through the duct network. Note that the supplemental resistance heat element is placed in this part of the system. At (5), the refrigerant is returned to the compressor and then to an expansion device, where it condenses the liquid refrigerant and returns it to the outdoor coil. The cycle repeats itself from this point.

The Cooling Cycle

During the summer months, a heat pump is designed to respond as an air-conditioning unit. For this to occur the reversing valve must be placed in the cooling-cycle position. In some systems this is accomplished by a manual changeover switch, whereas in others it is achieved automatically according to the thermostat setting. The operating position of the valve simply directs the flow path of the refrigerant.

When the cooling cycle is placed in operation it first causes the refrigerant to flow from the compressor into the indoor coil. During this part of the cycle the refrigerant is in a low-pressure gaseous state that is quite cool. As the circulation process continues, the indoor coil begins to absorb heat from the inside air of the building. Air passing over the indoor coil is cooled and circulated into the duct network for distribution throughout the building.

After leaving the indoor coil, the refrigerant must pass through the reversing valve and into the compressor. The compressor is responsible for increasing refrigerant pressure and circulating it into the outdoor coil. At this point of the cycle, the refrigerant gives up its heat to the outside air, is cooled, and is changed into a liquid state. It then returns to the compressor, where it is pumped through an expansion device and returned to the indoor coil. The process then repeats itself.

Figure 4-33 shows an illustration of the heat pump during its air-conditioning cycle. At (1), heat is absorbed from the inside air and cool air is transferred into the building. At (2), the pressure of the heat-laden refrigerant is increased by the compressor and cycled into the outside coil for transfer to the air. At (3), cool dehumidified air is circulated through the duct network as a result of passing through the cooled indoor coil. At (4), the refrigerant condenses back into a liquid as it circulates through the outdoor coil. At (5), the liquid refrigerant flows through the compressor and expansion device, where it is vaporized and returned to the indoor coil to complete the cycle.

STEAM AND HOT-WATER HEATING SYSTEMS

Steam and hot water are widely used today in comfort heating systems of buildings of moderate and large size. These systems are designed to use either electricity or fossil fuel as a primary source of heat to produce steam or hot water. The expanding pressure of steam forces it to pass naturally through the system, whereas hot water must be pumped through a network of pipes and radiators for distribution.

System radiators are then placed at strategic locations under windows or near doors, where the heat loss is greatest. Heat produced by this process is very dependable and provides a uniform method of distribution.

The systems concept that has been used with other comfort heating equipment can be applied equally well to steam and hot-water heating systems. The source of this system is often called a *boiler* because of the effect it creates when water is heated. The transmission path is somewhat unusual because it is formed by a network of pipes or tubes that is used to distribute steam or hot water. Control is usually accomplished by thermostats that sense the need for changes in heat and by valves that regulate flow to various rooms of the building.

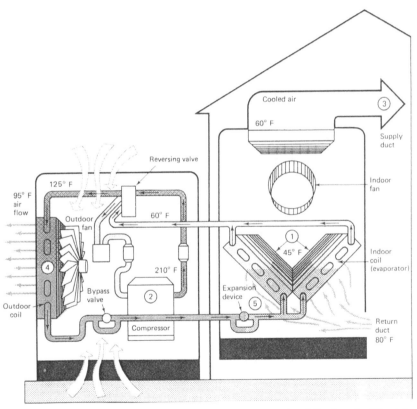

Figure 4-33. Cooling-cycle operation. (*Courtesy of Public Service Electric and Gas Company.***)**

Radiators serve as the system load and do work by giving off heat from a metal surface. Indicators may be used at a number of locations to monitor pressure and temperature. Figure 4-34 shows these components in a simplified steam or hot-water comfort heating system.

Fossil-Fuel Boilers

The boiler of a steam or hot-water system consists of a metal container that is heated by some type of a furnace assembly. In practice, coal, natural gas, fuel oil, or electricity is used as the primary source of heat energy. The applied energy causes water inside the container to become hot or to expand and change into steam. The capacity of the boiler depends largely upon its effective heating surface. In this regard, the surface area in contact with the water has the greatest effect on its capacity. To develop large surface areas for a given capacity, design is either of the *water-tube boiler* or *fire-tube boiler* type.

Water-tube boilers are designed to transfer heat to water that flows inside tubes that are placed in the combustion chamber. Heat gener-

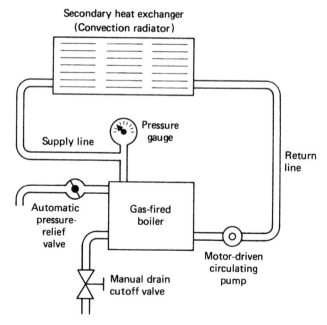

Figure 4-34. Components of a steam or hot-water comfort heating system.

ated by the energy source circulates over and around the outside of each water tube. The transfer of heat to water causes it to expand and produce a corresponding increase in pressure. An accumulation of hot water or steam at the top of each tube is eventually collected and distributed under pressure to the remainder of the system. Operational temperatures of a hot-water system range from 120 to 210°F (59 to 99°C). In steam systems, steam circulates at a temperature of 212°F (100°C), which generally requires operating temperatures slightly higher than those of a hot-water system. Figure 4-35 shows an internal view of a gas-fired water-tube boiler.

The immersion element of a *fire-tube boiler* is designed so that the heat source runs through the center of tubes that are surrounded by water. If fossil fuel is burned by the system, it is applied at one side of the tube and exhausted into a flue at the opposite side. Typical fire tubes for a gas-fired boiler are made with a 2-inch (5.1-cm) outside diameter of No. 13 gauge steel that is copper clad on the outside. Straight lengths of tubes are normally used to avoid traps or pockets for gas accumulations. Figure 4-36 shows a cross-sectional view of a fire tube energized by a gas burner. When installed in a boiler, a number of these tubes are arranged near the bottom of the water vessel. After being heated, water rises to the top of the boiler, where it is pumped or distributed under pressure to the remainder of the system. Figure 4-37 shows an assembled gas/oil-fired type of fire-tube boiler.

Electric Boilers

Electricity is frequently used today as the primary energy source for hot-water comfort heating boilers. Boilers of this type are usually factory-assembled and wired with a circulator pump, compression tank, and controls in a clean compact housing. Heat is produced by immersion resistance elements that are placed directly into the water vessel. Figure 4-38 shows a partial cutaway view of an electric hot-water boiler with the immersion elements exposed. Through sequential control procedures, different blades of the element are cycled on and off to assure even wear for prolonged operational periods. Through this procedure each element blade receives approximately the same operational time.

The immersion heating element of Figure 4-39 serves as the primary energy source of an electric boiler. Each blade of the element is

Figure 4-35. Inside and outside views of a 3-million Btu boiler

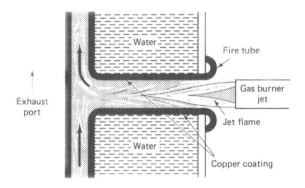

Figure 4-36. Fire-tube boiler construction.

Figure 4-37. Assembled gas/oil-fired hot-water boiler. (*Courtesy of The H.B. Smith Co.*)

Figure 4-38. Cutaway view of an electric hot-water boiler. (*Courtesy of Industrial Engineering Co.*)

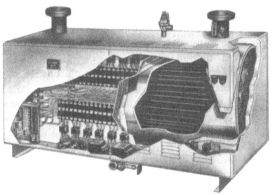

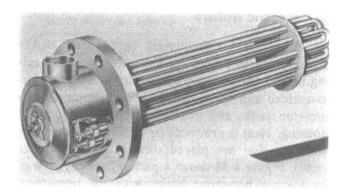

Figure 4-39. Immersion-type electric heating element. (*Courtesy of Patterson-Kelley Co.*)

made of stainless steel and is sheathed with a special alloy of indium and cobalt called Incoloy. This type of material does not interact with water supply minerals and provides smooth heat generation regardless of the electrical conductivity of the water. Element blades are connected through a moisture-resistant seal made of layers of epoxy and silicon rubber. This type of construction prevents condensation formation at the terminals which could cause electric arcing. Each blade of the heating element should be checked periodically for excessive wear to assure continuous operation at a high level of efficiency. Boilers that use this type of element react very quickly by bringing water temperature or steam up to an operating level in a very short period of time.

Transmission Path

The transmission of heat energy of hot water or steam is unique compared with the procedure followed by other systems. Steam or hot water, in a sense, serves as a medium which is circulated through a path from the source to the load device. Initially, the temperature of the medium is raised by the boiler. In a hot-water system, circulation of the medium is provided by a pump. A closed network of pipes serves as the transmission path between the boiler and the individual radiators. After heat is transferred to each radiator by the circulating water, it is returned to the boiler for recycling. This type of transmission path, which contains both a feed line and a return line, is considered to be a two-pipe or closed-loop system.

When steam is used as the transfer medium, it is possible to simplify the transmission path to a one-pipe system. The natural circulation of steam under pressure and the simplicity of the transmission path make this a very economical and desirable method of heating.

When water is heated by a boiler, it has a natural tendency to rise to the top of its container and cause a corresponding increase in pressure. These two conditions are utilized in the operation of the one-pipe transmission system. Steam pressure is quickly transferred through a network of pipes by rising to each radiator of the system (see Figure 4-40). After arriving at the radiator, heat energy is transferred to the metal parts of the radiator by the convection process. The release of heat by steam causes it to condense immediately or change into water. This water begins to accumulate at the bottom of the radiator, where it eventually flows back to the boiler through the supply line. Systems of this type are described as having positive or up feed with a gravity return of the condensate.

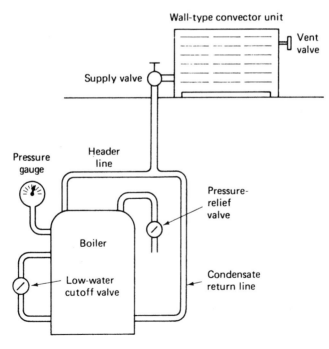

Figure 4-40. One-pipe heating system.

Heating Units

The heating units of a steam or hot-water system are placed under windows, near doors, in ceilings, or at the points of a room where the heat loss is greatest. This part of the system is primarily responsible for transforming steam or hot water into a useful form of comfort heat. Each unit does a certain amount of work when performing its function. A composite of all the work done by each unit is representative of the system's load.

The unit heater, commonly called a radiator, is composed of a set of tubes or pipes placed in a metal housing. In an ordinary steam or hot-water system, the radiator is often considered a secondary heat exchanger. It transfers the heat energy of steam or hot water into warm air. The shape and metal area of the radiator determines its heating capacity.

Figure 4-41 shows a horizontal-delivery heating unit with forced-air circulation capabilities. This unit is intended for use on steam systems operating at 30 psi or greater. Several models of this unit are available which range in capacity from 18,000 to 340,000 Btu

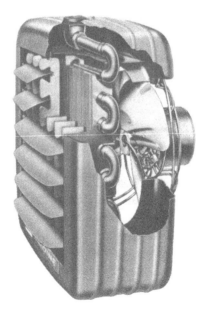

Figure 4-41. Horizontal-delivery hot-water heating unit. (*Courtesy of Modine Manufacturing Co.*)

per hour. Vertical fins attached to the radiating coil help to produce uniform heat transfer to the circulating air.

Vertical-delivery heating units are normally mounted in the ceiling and are designed to distribute air downward to a comfort zone at the floor level. This type of unit has forced-air circulation that can be selected from a variety of heat-throw patterns. Figure 4-42 shows a partial cutaway view of a representative unit. All air admitted to this unit must pass through the side-mounted aluminum fins bonded to the heating coils. The steam or hot-water passages are made of copper to assure maximum transfer of heat to air. Units of this type are designed primarily for hot-water or steam pressure not to exceed 30 psi. Heating capacity ranges from 42,000 to 610,000 Btu per hour for standard units and 33,000 to 470,000 Btu per hour for low-pressure applications.

Many of the radiators or heating units used in steam or hot-water systems do not have forced-air circulation capabilities. These units respond to the convection principle that has warm air expanding and rising from a surface after it has been heated. When this action takes place, cool air at the floor level moves in to take place of the upward flow of warm air. Through this airflow procedure, air continually circulates over the heated radiator coils. The amount of warm air developed by this type of unit is based upon the applied heat and the surface area of the heating element.

Figure 4-42. Cutaway view of a vertical-delivery hot-water heating unit. (Courtesy of Modine *Manufacturing Co.*)

Baseboard units such as the one shown in Figure 4-43 are frequently used for convection heating installations. These units are mounted at the floor level or under windows. When large rooms are to be heated, several baseboard units can be connected together and placed around the room. Comfort heating by this method has a high level of efficiency and is very reliable. The circulation of air from the lower level of a room to a higher level does not, however, produce an even distribution of warm air to all parts of a room.

INFRARED HEATING SYSTEMS

Any object that is hot enough to be seen in a dark room could be described as sending out visible rays or waves of light energy. At the same time, this object will be sending out a much greater number of heat waves. We cannot see these waves, but they can be felt and used to heat people and objects without changing the surrounding air. The term *infrared heating* is used to describe this action.

Infrared heating is a rather unusual form of energy that is used for "spot heating" or in supplemental units to improve the operational efficiency of other systems. In warehouses or buildings with high ceilings, convection heating is rather impractical. A great deal of energy is wasted getting heat to a specific location or to workers in a rather

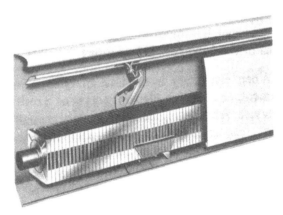

Figure 4-43. Baseboard convection heater. (*Courtesy of Weil McLain Inc., A Div. of the Marley Co.*)

limited area of the building. For applications of this type, infrared heating is used to transfer heat without the necessity of heating large volumes of air and circulating it throughout the building.

When a complete building or a local area is heated by the overhead infrared method, heat is released by an element through direct radiation of waves. This form of energy travels through the air at the speed of light. Objects such as walls, the floor, machinery, or workers located in the transmission path absorb the heat waves. Heat developed by each object is, in turn, released into the air by the convection process. This means that the solid objects of a building heated by infrared energy also serve as a secondary source of heat. It is important to remember that a person who works in an area of this type should be located between two or more overhead sources so that heat is received on both sides of the body.

Electric Infrared Equipment

When electricity is used to energize infrared equipment, the heating element serves as a resistive load device. Alternating current passing through the element resistance causes it to generate heat. In practice, the element becomes hot enough to produce a dull-red glow. Temperatures of 1450°F are generally needed to generate infrared energy of a wavelength that is readily absorbed by most solid objects. Figure 4-44 shows an overhead infrared heater with a metal-sheathed, single-ended heating element. Units of this type are available in capacities from 500 to 2500 watts for low-watt-density heating patterns.

The large mass of a metal-sheathed heating element of an in-

Figure 4-44. Overhead infrared heater. (*Courtesy of Space Ray, A Div. of Gas-Fired Products, Inc.*)

frared heater permits it to be controlled by alternately turning it on and off. Typical elements of this type take approximately 2 minutes or more to reach the correct color-radiating temperature. When it is turned off, it requires approximately the same amount of time to cool down. Because of its built-in time lag, this element can be cycled on and off to maintain a predetermined temperature.

A percentage input timer can be effectively used to control the on/off cycle of an infrared heating system. Controllers of this type are designed to have from 0 to 100% control for a 30-second operating cycle. If the timer is set at 60%, the elements would be energized for 18 seconds (60% of 30 seconds) and off for 12 seconds of each 30-second cycle. Manual adjustment of the operational cycle is set by the input percentage control knob of the timer. Heat-energy production is at a maximum value when the control is on continually or at 100% of the 30-second operational cycle. Lower percentage settings will produce lower heat-production values. Less electrical energy is consumed by the system when the timer control is set to a lower percentage value.

Gas Infrared Equipment

A large amount of the equipment used to develop infrared heat is energized by natural gas or bottled liquefied petroleum (LP) gas. This primary source of energy is essentially used to heat an infrared radiating element until it produces a dull red glow. Construction of the element and the unit is quite varied today. In some equipment a ceramic radiating element is heated by the combustion of gas. In other equipment a specialized burner is employed that produces radiation from tubular elements that are heated by the circulation of combustion gases that are drawn through the element by a vacuum blower. This type of equipment permits the interconnection of several individual units for heat distribution over large areas.

Gas-energized infrared heating units must be supplied with air to produce combustion and must exhaust unburned gases into the atmosphere. The ignition of gas for this type of unit is primarily achieved by an electronic pilot spark. This part of the system is very similar in operation to that of a gas-fired comfort heating furnace. Figure 4-45 shows an example of a radiant-tube infrared heater. The burner manifold, gas control, ignition unit, safety components, and blower assembly are all enclosed in the steel cabinet. The radiating

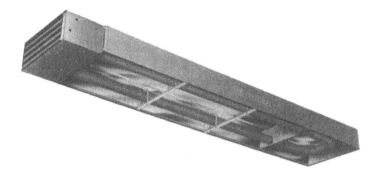

Figure 4-45. Radiant-tube gas-fired infrared heater. (*Courtesy of Space Ray A Div. of Gas-Fired Products, Inc.***)**

element is a U-shaped tube made of aluminized steel tubing that has approximately 25 square feet of surface area. The operating temperature of the element surface must reach 950°F to produce infrared radiation at an efficient level. A number of individual units may be connected together and serviced by a common gas source.

ENERGY CONSERVATION CHECKLIST
FOR HEATING SYSTEMS

The heating system is responsible for a very high percentage of the building's total energy consumption. The rising cost of energy and the demand for its conservation has caused a great deal of concern to be expressed for better energy management. General equipment operational procedures and system efficiency are the primary areas of concern.

A heating system checklist is presented in Table 4-1 as a means of initiating some rather important energy conservation measures. The list is rather general so that it can be applied to nearly any of the major methods of heating. Implementation of some measures may involve adjustments, calibration, and equipment modification. Do not attempt to make these changes without first consulting the appropriate equipment operational manuals or engaging professional assistance.

Table 4-1: Checklist for Heating Systems/Saving Natural Resources

Items to check	Corrective action
General Operational Procedures	
Check the temperature setting of each thermostat.	Adjust the temperature setting to 68°F when occupied or 60°F when unoccupied. Interior offices tend to experience heat gains when people are present, lighting is used, and when some equipment is placed in operation.
Evaluate building heating-time schedule.	Reduce heating when buildings are unoccupied. Preheat building so that it reaches 65°F by the time occupants arrive. Reduce heat during the last hour of building occupancy. Turn off all portable or unit heaters when the building is unoccupied. Reduce heating to dock areas, storage rooms, garages, and platform areas.
Evaluate the natural humidity of the building.	Measure the RH (relative humidity) of the building and adjust humidifier operation so that a natural range of 20 to 60% RH occurs before humidification equipment is energized.

Table 4-1: Checklist for Heating Systems/Saving Natural Resources (Continued)

Items to check	Corrective action
General Operational Procedures	
Investigate building heating practices for different areas.	Reduce or turn off heat to lobbies, corridors, vestibules, stairways, or large storage spaces.
Equipment Operational Procedures	
Check the operational time of heating equipment with respect to outside temperature. Must the heating system operate continually, or can it be controlled automatically for operational cycles?	If boilers or heaters operate continually, the system may need to be modified to operate more efficiently. Prepare an operational sequence study and make recommendations for changes if necessary.
Test the water temperature of a hot-water system with respect to outside temperatures.	Consult with the building engineer for appropriate hot-water operating temperature. It may be possible to reduce hot-water temperature without causing a radical change in system operation. Study how automatic temperature controls respond to outside temperature changes. Investigate the possibility of changing the operating range automatic temperature controls.
Check the heating system during summer operation.	If boilers or furnaces are not in use, it may be possible to turn off the pilot lights. If pilot replacement is needed, investigate the possibility of using electronic ignition.

Table 4-1: Checklist for Heating Systems/Saving Natural Resources (*Continued*)

Items to check	Corrective action
Equipment Operational Procedures	
Study the flue and stack emission of fossil-fuel-burning systems for excessive smoke and soot production.	With the building engineer or qualified personnel, inspect burner nozzles for wear, dirt, or incorrect spray angle. Perform flue-gas analysis and set the fuel/air ratio to a proper value.
Investigate boiler and furnace general maintenance practices.	Make a general maintenance plan for the heating system. It should include tests to determine if scale-deposit removal is necessary or if there is an accumulation of sediment in water tubes. The plan should call for periodic inspection of the flue and fire tubes, burner nozzles, boiler insulation, brick work, and casings for hot spots and air leaks.
Evaluate the operational cycle of the burner assembly.	If the burner short-cycles, investigate the limit switch setting and the thermostat.

Table 4-1: Checklist for Heating Systems/Saving Natural Resources (*Continued*)

Items to check	Corrective action
Equipment Operational Procedures	
Check the temperature of the air in various rooms or areas of the building.	Measure air temperature with an accurate heating system thermometer and compare measured values with thermostat settings. With professional assistance, recalibrate the thermostat if necessary, clean control lines, recalibrate controllers, check the heat distribution ducts for obstructions, and make certain that heat-radiating surfaces are free of dirt and accumulated dust.
Evaluate the operation and effectiveness of hot-water radiators or convectors for proper output.	Trace the steam or hot-water lines to determine if the radiating elements are responding evenly. Check the return line for excessive heat to determine if traps are operating properly. Clean and replace thermostatic control valves. Check for radiator air locks and air grille obstructions.
Investigate the condition of steam-line or hot-water pipe insulation.	Trace the heat distribution network for broken or missing insulation. Repair or replace as needed. Large savings can be achieved when proper insulation is installed and maintained.

Table 4-1: Checklist for Heating Systems/Saving Natural Resources (*Continued*)

Items to check	Corrective action
Equipment Operational Procedures	
Evaluate the airflow duct network.	Establish a regular maintenance plan for replacing air filters,cleaning radiator surfaces, and checking air flow through the duct network.
Check the electrical rotating machinery part of the heating system.	Establish a regular maintenance plan for lubricating bearings and testing belt tension, alignment, and condition. Evaluate motor operating temperature and electrical energy consumption.

Chapter 5

Summer Air-Conditioning Systems/ Saving Natural Resources

INTRODUCTION

At one time, the inside of a building could be adjusted to meet human needs only by supplying heat in the winter and by opening the windows or using the circulating fans to move air in the summer. The cooling of air in the summer was considered to be a luxury that could be enjoyed by only a few. Since the early 1950s, most new industrial, commercial, and residential buildings have been equipped with summer air-conditioning units. Summer air-conditioning is now considered to be a necessity in many buildings and is an essential part of our modern society.

The term air conditioning generally implies a number of things other than just cooling the inside air of a building. The total comfort of the human beings plus the various equipment occupying a building must be taken into account. This involves such things as temperature control, moisture content of the air, air purity, and air movement. These factors must all be controlled on a continuous basis to have total comfort.

Equipment that provides total air-conditioning comfort for the occupants is generally called a complete air-conditioning system. This type of system must be capable of heating, cooling, purifying, humidifying, dehumidifying, ventilating, and circulating air into selected spaces. It is rather difficult to find one system that performs all these functions effectively. The heat pump discussed in Chapter 4 is the best example of a complete airc-onditioning system. The operational efficiency of the heat pump, with its cooling and heating capabilities being achieved by the same equipment, makes it a very popular air-conditioning system.

135

A large part of the summer air-conditioning equipment in operation today is not classified as a complete air-conditioning system. Heating and humidification, for example, are often accomplished by entirely different equipment. The comfort heating system is primarily responsible for this function. Summer air-conditioning equipment is usually independent of the heating system and is often considered to be a supplementary system. The duct network and air-moving equipment of the heating system may be used by the summer air-conditioning system to avoid a needless duplication of equipment.

AIR-CONDTIONING-SYSTEM CLASSIFICATIONS

There are three rather general equipment classifications of summer air-conditioning systems. These are largely determined by the size and design of the structure, equipment assembly procedure, the location of equipment, and the intended function of the building. Included in this classification are central-station systems, unitary systems, and combinational systems.

Central-station systems generally represent the most costly and massive form of air-conditioning equipment in operation today. This type of installation is usually designed by an engineer, purchased directly from the manufacturer, and erected at the building when it is being constructed. A central equipment room or mechanical room houses the system, and the distribution of conditioned air or cool water is through sheet-metal ducts or pipes. The size of this type of system starts at 25 to 50 tons and extends up to several-thousand-ton units.

Central-station systems operate by transferring a heated or cooling medium to individual space air terminals. In practice, water is a very common liquid medium. A number of unique components are needed in the physical makeup of the system to make it operational. These include such things as chillers, air handlers, water towers, and controllers, which become quite complex in terms of the on-site installation. Figure 5-1 shows the mechanical room of a building equipped with a central station heating-cooling system. Hot or cold water is pumped to individual room units in or near the spaces to be cooled.

Figure 5-1. Central-station mechanical room used in the past. (Many modern systems are decentralized.) (*Courtesy of Carrier Corporation*)

Unitary systems represent the most popular form of air conditioning for small- to medium-size buildings today. These systems are factory assembled, and tested and usually come in a self-contained package that is easily installed. An entire building may be served by a unitary system through a suitable duct network, or multiple units may be installed to serve specific areas. Typical installations have a large part of the equipment located outside the conditioned space. Rooftop-mounted units, wall assemblies, and ground-level installations are quite common. Figure 5-2 shows a representative roof-mounted unitary air-conditioning system.

Combinational air-conditioning systems have the features of a central station and a unitary system in a smaller, more compact assembly. This type of system is generally placed in a central mechanical room that is conveniently located for ease of air distribution. Heat energy is supplied to air-handling units in the form of steam or hot water. Chilled water is distributed from a central refrigerating unit over the same network of pipes. Individual cooling units may also be employed to condition air for specific areas, or specialized zoning equipment may be used to direct air to individual rooms. Operation of the

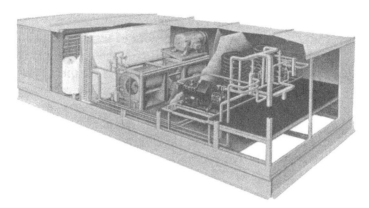

Figure 5-2. Roof-mounted unitary air conditioner.

system is regulated by thermostatic controls located in each room or area served by the system.

Mechanical Refrigeration

Mechanical refrigeration is the operational basis of nearly all summer air-conditioning systems. This part of the system is primarily responsible for removing heat from a selected space and transferring it outside or into another area. The fundamental principle of operation is based upon the *second law of thermodynamics*. This law states that heat can pass only from a warm body to a colder body. The flow of heat from a warmer body to something colder is generally called *heat transfer*. Mechanical refrigeration achieves heat transfer by circulating a refrigerant through pipes or coils into an area where cooling is desired. Heat is transferred from the air of a selected space to the circulating refrigerant. Through this action the remaining air will be of a lower temperature.

AIR-CONDITIONING SYSTEMS

Summer air conditioning can be accomplished today by a variety of rather basic but fundamentally different methods of cooling:

1. Evaporative cooling
2. Cold-water cooling
3. Steam-jet cooling

4. Gas compression refrigeration
5. Absorption refrigeration
6. Thermoelectric refrigeration

As noted, the first three methods of air conditioning are described as cooling processes. These methods are used exclusively to achieve air-conditioning applications in general, and are somewhat limited for a number of reasons. Evaporative cooling, for example, is restricted primarily to dry or low-humidity regions of the country. By comparison, cold-water cooling is effective only where there is a continuous source of cool water. Steam jet cooling is by far the most complex of the three methods. It is used primarily in large commercial and industrial applications.

The last three methods of cooling are described as refrigeration processes. Essentially, these methods could be used in commercial and home refrigeration equipment as well as air conditioning. In general, rather low temperatures are produced by refrigeration equipment. A very high percentage of our residential and commercial building air-conditioning systems utilize the compression principle of operation. Absorption refrigeration is somewhat more complex than the other methods and is used primarily in large industrial installations. Thermoelectric refrigeration is the newest method of air conditioning. Cooling is produced by passing an electric current through several thermocouples connected in series.

Evaporative Cooling Systems

An evaporative cooling system is designed to cool the inside air of a building or space by lowering its dry-bulb temperature. Essentially, this is accomplished through the use of water evaporation. Water is either sprayed into the air or used to saturate a fibrous pad. Air is cooled by forcing it to pass through the pad or water spray. Figure 5-3 shows a fibrous pad evaporator with a drip-type water distribution system.

The essential parts of an evaporative cooler are the blower, blower motor, water pump, water distribution tubes, water pads, and a cabinet with louvered sides. In operation, the blower assembly draws outside air through the louvers, where it comes in contact with the moisture-laden water pads. Air passing through the unit is then

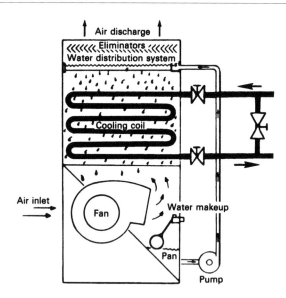

Figure 5-3. Drip-type evaporative cooler.

distributed by a duct network to selected rooms or areas.

Water held in the pads of the evaporator is used to absorb heat from the air as it passes through the louvers. This action causes a certain amount of the water to evaporate, which in turn lowers the drybulb temperature. Water consumed by evaporation is returned to the pads by pumping action through the distribution tubes.

Evaporative cooling systems are unique compared with other methods of cooling. The evaporator unit is usually located outside the building where cooling is to take place. Dry outside air is pulled into the evaporator and cooled before being forced into the building. New air is continually drawn into the unit during its operational cycle. In practice, evaporative cooling systems are not very effective in humid climates.

Cold-Water Cooling Systems

Summer air conditioning can be achieved in many localities by passing warm room air over a water-cooled coil and circulating the air back into the room. Systems that respond to this method of cooling require a natural or artificial source of cool water. Figure 5-4 shows a simplification of the cold-water cooling system.

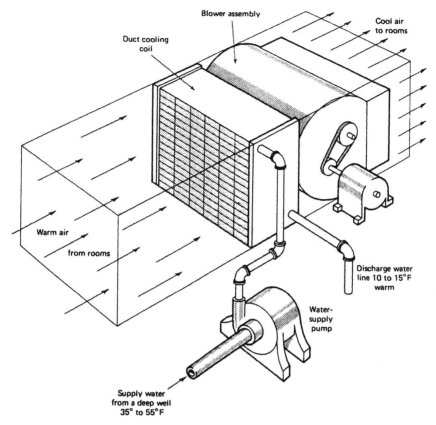

Figure 5-4. Cold-water cooling system.

The essential parts of a cold-water cooling system are the supply water pump, the duct cooling coil, and a water discharge sump. In operation, cool water of 35 to 55°F or 1 to 13°C is forced into the system by the water supply pump. This water is then circulated through the duct cooling coil. Warm air circulating around outside metal fins of the cooling coil is chilled and passed into the duct for distribution. Heat absorbed by water flowing through the cooling coil is discharged into a sump or dry well. Discharge water temperature is increased by approximately 10 to 15°F (–12 to –9°C). Cold-water cooling systems are not very effective in many areas of the country because of the high temperature of the water supply source.

Steam-Jet Cooling Systems

Steam-jet cooling is based on an operational principle that deals with the relationship of pressure and the boiling point of a liquid. A reduction in pressure, to which a liquid is subjected, causes its boiling point to be lowered. As water boils, it evaporates very rapidly. Evaporation causes the temperature of a liquid to be reduced. When liquid of a certain temperature is sprayed into a closed low-pressure vessel, it enters the vessel at a temperature higher than its boiling point will be at the reduced pressure. This action causes immediate evaporation of the liquid.

The components of a steam-jet cooling system are shown in Figure 5-5. In operation, water returned from the space or room cooling units is sprayed into the round evaporator tank. The partial vacuum of this vessel causes immediate vaporization of a high percentage of the sprayed liquid. Unevaporated water falls to the bottom of the evaporator tank at a temperature of 40°F (4°C). At this temperature it is pumped from the evaporator tank and recirculated to the room cooling units to again absorb heat for completion of the operational cycle.

The partial vacuum of the evaporator is created and maintained by a jet of high-pressure steam passing through the booster ejector. This jet creates suction within the space above the water of the evaporator and lowers the pressure of the chamber. Evaporation of the return water absorbs heat from the returned liquid and passes through the booster ejector. Vapor and steam from the ejector is then applied to the booster condenser. System water circulating through the condenser tubes causes the steam applied to the outside of the tubes to change into water. The condensate pump removes this water from the condenser. A smaller auxiliary condenser pulls air from the booster condenser for further condensate removal before it is returned to the steam source.

Steam-jet cooling systems produce very little noise or vibration during operation, occupy very little space, and have practically no moving parts other than the pumps. They require a constant source of steam when in operation. This could be derived from the operation of other machinery or an independent steam boiler. Systems of this type are used primarily to cool large commercial buildings and industrial facilities.

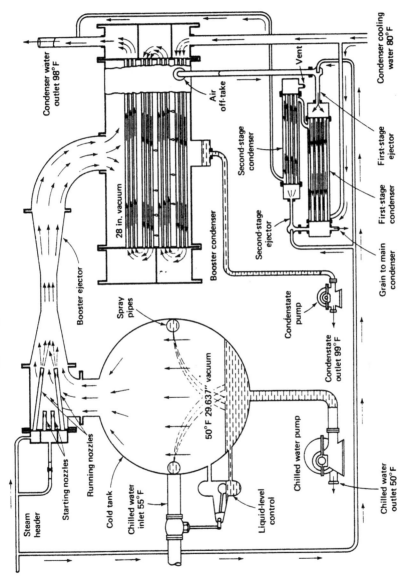

Figure 5-5. Steam-jet cooling system. (Courtesy of ASHRAE)

Gas Compression Refrigeration Systems

Compression refrigeration is a process through which a refrigerant is mechanically circulated through a closed network of components. The refrigerant is alternately compressed, liquefied, expanded, and evaporated during the operational cycle. Every component of the system plays a specific role in the process of removing heat from the space to be cooled and transferring it to a place where its presence is not objectionable.

The components of a compression refrigeration system are shown in Figure 5-6. The essential parts of this system are the evaporator coil, compressor, condenser, liquid receiver, and flow control device. During a normal operational cycle, air from the space being cooled circulates around the outside of the evaporator coil. Heat from this air is transferred to the cooler *evaporator* coil. Liquid refrigerant flowing inside the evaporator absorbs heat from outside air and causes the refrigerant to boil and be carried away as a gas or vapor. The size, shape, and design of the evaporator is determined by the capacity of the air conditioner. This part of the system may be located in the space being cooled, or in a duct network. In most installations the evaporator is located a considerable distance away from the other components of the system.

The *compressor* is essentially a pump that is used to remove vapor from the evaporator coil. This part of the cycle takes place during the suction stroke of the compressor. Pulling the refrigerant vapor from the evaporator is necessary to keep the evaporator pressure and coil temperature at a low value. The suction stroke of the compressor generally has a rather low operating pressure.

After the refrigerant vapor has been drawn into the compressor during the suction stroke, it is squeezed together during the compression stroke. Vapor compression is used to increase the gas temperature to a value that is high enough to permit heat rejection of the vapor when it reaches the condenser. Additional heat produced by compression of the hot refrigerant vapor is called *superheating*.

Heat rejection of the superheated refrigerant vapor takes place in the *condenser* of an air-conditioning system. In this component, hot refrigerant vapor is cooled by lower-temperature air or cool water passing over the outside surface of the condenser coil. Heat of the refrigerant vapor is absorbed by the outside air, thus reducing the vapor

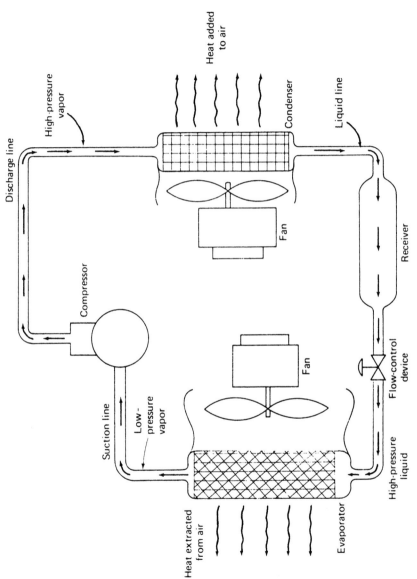

Figure 5-6. Components of a compression refrigeration system.

temperature. This action, in turn, causes the vapor to condense into a warm liquid that collects in the liquid receiver.

The *liquid receiver* of the system is primarily a storage tank for the pressurized liquid refrigerant. This tank is on the outlet side of the condenser, which is on the high-pressure side of the system. In addition to its storage function, the receiver strains the refrigerant to remove scale, dirt, and moisture. Refrigerant stored in the receiver is eventually transferred to the compressor through the flow control device, where the cycle is repeated.

The *flow control device* of the system is designed to automatically regulate the flow of liquid refrigerant between the condenser and the evaporator coil. The quantity of refrigerant supplied to the evaporator is determined by the system temperature and compressor operational efficiency.

In air conditioners, flow control is frequently achieved by an expansion valve. This type of valve is usually operated by a bellows arrangement containing a gas. The pressure of the control gas is determined by evaporator temperature. When the evaporator temperature is low, the gas pressure in the thermal sensor bulb and bellows is low. This action causes the valve to close, which prevents refrigerant from entering the evaporator. An increase in evaporator temperature causes an increase in sensor bulb temperature and a corresponding increase in bellows pressure. This, in turn, causes the expansion valve to open and permits the refrigerant to enter the condenser, where it adds to the cooling process. In a sense, the expansion valve simply alters the flow of refrigerant to the evaporator according to the temperature as determined by the sensor bulb. A cutaway view of the expansion valve is shown in Figure 5-7.

Effects of Heat Transfer

Heat transfer produces several rather unusual effects when it occurs in an air-conditioning system. The first of these deals with the temperature of the body or substance involved in the process. In air conditioning, our primary concern is with the temperature of the air in a selected room or space. Through the heat-transfer process air becomes cool through a lowering of its temperature. For this to take place, heat in the air must be absorbed by a chilled refrigerant. This action causes the temperature of the air to be reduced and the refrigerant to become warmer. Through this process, it is possible to maintain the temperature of a room at a desired level.

The physical state of a refrigerant is influenced to some extent by the heat-transfer process. Lowering the temperature of a gaseous refrigerant will, for example, cause it to change into a liquid. This action is commonly called *condensation*. Gases have a natural tendency to lose heat when they condense.

The *condenser* of a mechanical refrigeration unit is similar in appearance and function to the radiator of an automobile or the heating element of a hot-water comfort heating system. Air forced to pass

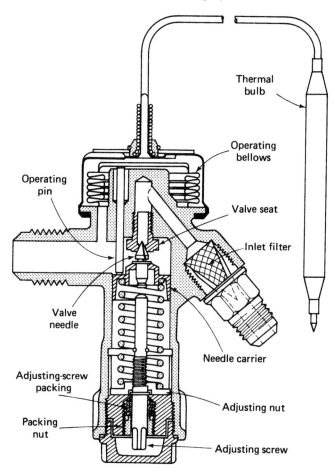

Figure 5-7. Cutaway view of an expansion valve.

around the outside of the condenser coil assembly removes heat from a gaseous refrigerant that circulates through the inside of the coil. Lowering the temperature of the gas eventually causes it to change into a cool liquid that is circulated through the remainder of the system.

If sufficient heat is added to a refrigerant when it is in a liquid state, it will change into a vapor or gas. The term *vaporization* is normally used to describe this action. Liquids have a tendency to absorb a great deal of heat when they vaporize. The temperature at which vaporization takes place is generally called the *boiling point*. Water at normal atmospheric pressure will boil at 212°F (100°C). An air-conditioner refrigerant, such as DuPont™ Freon-12, boils under normal atmospheric pressure at a temperature of –21.6°F (-30°C). The boiling point of a refrigerant changes a great deal with pressure. DuPont™ Freon 12 at a gauge pressure of 30 psi or 30 psig will vaporize at 32°F (0°C). Heat from a room will be readily absorbed by the refrigerant and cause it to vaporize.

In an air conditioner, vaporization takes place in a finned coil structure similar to the automobile radiator. This part of the system is called an *evaporator*. Evaporation describes the process whereby vapors escape from liquids or fluids. Warm air circulating through an evaporator causes the liquid refrigerant to boil and change into vapor. Heat absorbed by this action is then transferred away from the area by the vaporized refrigerant.

Reclamation of Refrigerant

The most common refrigerant, a chlorofluorocarbon gas, is presently being stipulated by the U.S. government to be reclaimed and recycled instead of released into the atmosphere when it is drained from any compressor. The reasoning behind this law concerns itself with the environment. Chlorofluorocarbon gasses, once released into the atmosphere, pollutes the air. The law, accompanied by a hefty fine for any violators, states that once a company drains the gas, it must then be reclaimed into a specially designed tank. The tank must then be transported to a designated reclamation area where it is collected and filtered. The original company receives a certain amount for the reclaimed gas. The gas is then filtered again and sold to manufacturer's to use again as a refrigerant in a compressor.

Absorption Refrigeration

The absorption refrigeration method of cooling is unusual to the extent that it uses heat rather than electricity as a source of energy. This heat can be derived from a gas- or oil-fired steam generator, or it may be developed directly by a gas- or oil-fired burner. In most systems water is used as the refrigerant, and lithium bromide serves as the absorbent. The system's absorption unit operates under a vacuum which provides the refrigerant with a boiling point low enough to achieve comfort cooling.

A diagram of an absorption refrigeration system is shown in Figure 5-8. Essential parts of the system are the evaporator, absorber, generator, separator, and condenser. Installations of this type appear to be somewhat simplified when compared with other methods of refrigeration. Normally, there are no moving parts in the internal structure of the system. Its complexity lies in the network of internal valves and controls that are needed to regulate the flow of the refrigerant.

During the refrigeration cycle, heat from a steam jet or a name is applied to the generator. This action causes a water/lithium bromide mixture to begin boiling. A percolator like structure inside the generator is used to force the resulting vapor up into the separator.

The *separator* is basically a receiver tank that has one inlet and two outlet lines. Since lithium bromide evaporates at a much lower temperature than water, the two divide immediately upon entering the separator. Water vapor rises and flows into the condenser. Lithium bromide cools somewhat and condenses into a liquid. An accumulation of liquid lithium bromide appears at the bottom of the separator tank. Eventually, this solution flows from the separator tank to the absorber.

Water vapor rising from the top of the separator tank is applied to the *condenser*. Upon entering the condenser, the vapor is cooled by either air or water circulating over its outside surface. This action causes the vapor to change immediately into a liquid. The water refrigerant is then gravity-fed through a tube to the top of the evaporator coil.

The *evaporator* of the system is primarily responsible for the transfer of heat from the room or space being cooled. Water from the condenser and hydrogen gas from the absorber are mixed together in the condenser. Hydrogen is used to equalize pressure between the condenser and the evaporator. This liquid-gas combination then absorbs heat from warm air circulating over its outside surface. The

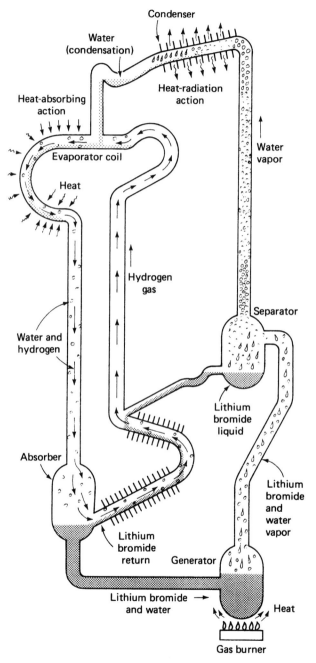

Figure 5-8. Absorption refrigeration system.

presence of hydrogen in the mixture speeds up the boiling or vaporization process. Vapor produced by the mixture is then responsible for carrying heat away from the evaporator. A pipe connected to the evaporator provides a path for the flow of vapor to the absorber.

In the *absorber,* a solution of lithium bromide takes in the vapor solution applied to it from the evaporator. The absorber solution is derived from the separator after it boils out of the generator. Lithium bromide in either liquid or dry form has a very strong attraction for water vapor. The absorption rate is generally increased at lower temperatures. A water-cooling coil is usually included in the outer shell of the unit to increase its absorbing action. Completion of the cycle occurs when the resulting absorbent mixture and refrigerant returns to the generator for the start of a new cycle. Hydrogen is also released by the absorber when water vapor is absorbed by lithium bromide.

An absorption liquid-chiller assembly is shown in Figure 5-9. Units of this type can be selected in a capacity range from 120 to 1377 nominal tons of cooling. Thermal-gain options are also available for energy-saving operation. Absorption refrigeration equipment is used primarily in large-capacity air-conditioning systems.

Thermoelectric Refrigeration

Thermoelectric refrigeration is a rather new method of cooling that is presently being used in low-capacity air-conditioning systems. This method of cooling has been used for a number of years in small refrigerator units. Cooling does not involve moving parts or the flow of a liquid or gas.

Thermoelectric refrigeration is achieved by forcing an electric current to flow through a series of thermocouples. A thermocouple is essentially a bonded junction of two dissimilar materials. In a sense, the flow of electrical current can be considered as the working substance of the cooling system. When two junctions are connected together in series, a direct current (DC) will flow through the circuit. Each electron of the current will carry heat away from one junction and deliver it to the other junction, where it is given off. As a result of this action, one junction becomes cold and the other becomes warm. A special class of materials known as *semiconductors* are used in the construction of the thermocouples. Bismuth telluride is a common material used in the construction of these devices.

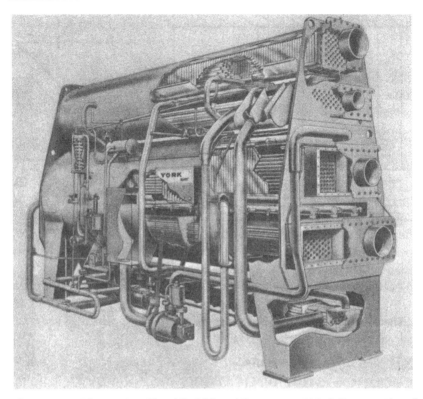

Figure 5-9. Absorption liquid chiller. (*Courtesy of York International Corp.*)

A simplification of the thermoelectric refrigeration system is shown in Figure 5-10. In operation, DC power is applied to a number of thermocouples connected in series. Heat is transferred from one side to the other, depending upon the direction of current flow. Warm air from the space being cooled is forced to circulate over the metal fins of each hot junction, then is removed from the system and rejected to a place where it is not objectionable. Reversing the direction of current flow through the circuit will cause it to generate heat and reject cool air. As a result of this characteristic, thermoelectric units are capable of producing year-round air conditioning on a limited basis.

One type of thermoelectric refrigeration that has gained in popularity is the thermoelectric module or Peltier device. These small, solid state devices are compared to heat pumps because of their abil-

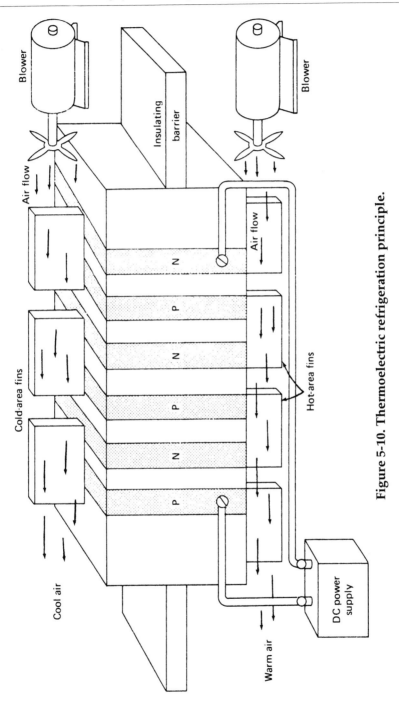

Figure 5-10. Thermoelectric refrigeration principle.

ity to produce heating or cooling effects. While typically somewhat small (a few square inches), they can be 'stacked' to increase capacity, but currently are not capable of cooling an entire room. Because of their ability to produce heat or a cooling effect, they are become popularized in portable travel coolers capable of keeping foodstuffs warm or cool during a journey. Once plugged into a DC power source, the modules can produce enough cooling effort to keep drinks and snacks cool. Conversely, if the flow of current is reversed, the same portable travel cooler can now be used to keep foodstuffs warm over a period of time. While not efficient because of the large current draw required, thermoelectric devices do offer other advantages over typical mechanical gas compression systems because they require no mechanical components. Therefore, thermoelectric refrigeration systems are relatively maintenance free and have a long service life without the hazards of refrigerants.

COOLING-SYSTEM APPLICATIONS

The application of cooling equipment to the air-conditioning system of a building structure can be achieved in a variety of ways. Primarily, such things as type of construction, floor plan of the structure, function of the building, and method of cooling must all be taken into account. More important, however, is the type of building heating system. Central heating systems tend to be a predominate consideration in this regard. Forced-air circulation systems and liquid or steam distribution through pipes are the major types of distribution to take into account. Cooling equipment can be applied equally well to either type of distribution system.

Some of the common types of cooling-system applications are
1. Water chillers
2. Independent system cooling
3. Year-round air conditioning
4. Central cooling packages

Water Chillers

Water-chilling applications are used primarily to add cooling and dehumidification to a steam or hot-water heating system. In a sense, most water chillers are mechanical refrigerator units of the compression type. Chillers of this type are effectively used for the cooling of small- to medium-size buildings. A compressor, condenser, thermal expansion control valve, and evaporator coil are included in the basic components of the chiller. Essentially, water is cooled in an evaporator coil and pumped into the system through the same pipe network that is used to distribute steam or hot water.

A representative water-chiller cooling system is shown in Figure 5-11. The chiller unit and boiler are completely independent of each other. Separate circulating pumps are employed by each unit. Cool water from the chiller is circulated through the steam or hot-water lines during the cooling season. Hot water or steam is circulated through the same network during the heating season. A changeover

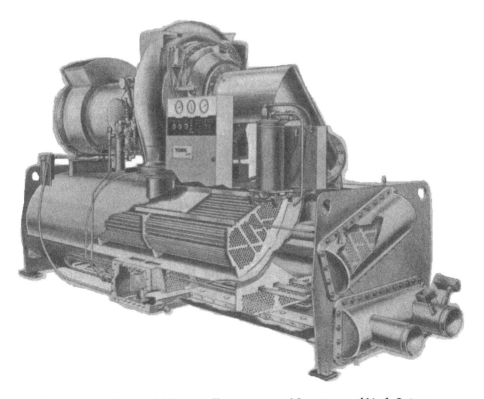

Figure 5-11. Water-chiller cooling system. (*Courtesy of York International Corp.*)

valve is generally employed to actuate the appropriate section of the system according to the demands of the season.

Convector units of the system are placed at strategic locations in each room of the structure. In large systems, each convector may employ a circulating fan and air filter. Independent control of the fan and cool-water flow is generally included. In small building installations a central control thermostat may be used to regulate the temperature of the entire structure. Additional examples of chillers can be seen in Figures 5-33, 5-34, and 5-35.

Since the same pipe network of the system is used for the distribution of hot or cool water, it must be completely insulated. During the cooling season, warm air has a tendency to condense on the cool pipes. Unless this action is minimized, condensate will accumulate throughout the system. In many installations special condensate drain networks are installed to alleviate this problem.

Water chillers may be entirely independent of the boiler or may be included as a part of the complete heating-cooling package. Independent chillers are generally used when cooling is added to an existing heating system.

Cooling Tower and Chiller Systems

While chillers are used to supply chilled water to a heat exchanger or convection unit located in an air handler to aid in cooling, cooling towers are devices used to remove the heat from the chiller system in a cooling cycle. While the term cooling tower is very general and used to describe any unit that capable of removing heat, including large scale units such as those found alongside power generating plants, this unit will focus on the scale and types of cooling towers used in heating and cooling systems such as shown in Figure 5-12. In general, these towers contain only a few major components. Because the primary purpose of the cooling tower is to remove heat from a system, the central component is a heat exchanger. While some towers use only convection to move air, most towers in air cooling systems use some method to move ambient air over the heat exchanger, such as a fan. The remaining components include a structure to contain the components and a common loop connecting the chiller to the cooling tower. This loop typically contains water, however this water is often filtered and treated for various contaminants both particulate and organic. While some chiller systems are designed to run year-a-round, many do not operate during heating seasons. In times when cooling is not needed, chillers and cooling towers may be shut down

Inside the cooling tower, various forms of heat exchangers are used to remove the heat from the water in the common loop and discharge it. While this may be a dry exchange or a wet, evaporative exchange, the tower must have a method for removing the heat from the loop. To supplement the heat exchange, a constant flow of air must be made through the cooling tower. Most systems use a fan to either draw the air (induced draft) or push the air (forced draft) through the system. A large 30 horsepower motor and fan assembly for the cooling tower shown in Figure 5-12 is shown in Figure 5-13. Advanced cooling tower systems can detect the amount of heat energy load needing to be dissipated. When the load is light, they operate in convection (draft) mode, and when the load on the unit is too great, the fans are used to supplement the flow of air over the heat exchanger coils.

Figure 5-12. Cooling Towers

Figure 5-13. Cooling Tower Fan Assembly

Independent Cooling Systems

An independent cooling system is considered to be completely unrelated to the heating system. In a hot-water or steam system, heat is distributed through a common network of pipes or tubes. The cool-air part of the system, being unrelated, must be distributed through its own network of air ducts feeding each room or space. Cooling by this method is generally achieved by a forced-air system, and heating is through a closed network of pipes.

An independent cooling system is generally somewhat more costly to install than other methods of cooling. This is due largely to the necessity for some duplication of equipment. A duct network, for example, must be connected to each room or space being cooled. Condensation buildup in the duct may also necessitate additional insulation and possibly a drain network. Circulation fans and blowers are also needed for both heating and cooling units.

Year-Round Air Conditioning

Year-round air conditioning is designed to produce heating and cooling through a combined unit housed in a single cabinet. The combined unit is called upon to heat, humidify, and filter the air during winter operation while cooling, dehumidifying, and filtering the air during the summer. Systems of this type usually employ a common blower or duct network for the distribution of air. Operational units may employ an air-cooled, water-cooled, or evaporative type of condenser. The distribution of air is largely dependent upon the type of condenser employed by the system.

Year-round air-conditioning units are primarily designed for high operational efficiency in a compact space. Figure 5-14 shows a representative terminal unit of the incremental type. This particular system is designed for new construction in motels, hotels, schools, offices, apartments, and so on, where one-room or small-area conditioning is needed. With this type of system, each room is provided with individual control.

Construction of an incremental unit is usually quite compact. Facing the conditioned space is the room cabinet, in which the discharge and air-return functions are provided. All mechanical equipment is housed in the cabinet assembly. The control console on the right side is responsible for fan speed and temperature settings. This particular unit employs a small energy-efficient heat-pump assembly. Wall sleeve mounting provides a seal for outside moisture and air infiltration. Cooling capacity ranges from 6500 to 13,800 Btu per hour with a heating capacity of 6400 to 14,000 Btu per hour for incremental

Figure 5-14. Incremental year-round air-conditioning unit. (*Courtesy of The Singer Company.*)

units of this type.

Year-round air conditioning can also be produced by a single unit and distributed through ducts to individual rooms or spaces. In single-zone units, air is distributed through ducts to different rooms or spaces. Figure 5-15 shows a cross-sectional view of a rooftop installation. Note that a conventional trunk duct is used for the supply and return air that goes down into a dropped ceiling for distribution. It is then routed to branch ducts and air diffusers according to the layout of the system. Airflow is generally the same to all areas of the system. A single-package gas-heating and electric-cooling unit is shown in Figure 5-16.

Multizone year-round air conditioning is another type of system found in larger industrial and residential buildings. In this system independent ducts are used to distribute conditioned air to different rooms or spaces. Some manufacturers prefer to use constant air circulation and temper the heating, cooling, and humidity levels according to the needs of a particular zone. Other manufacturers use a damper arrangement to premix the air before it enters each zone. Still others use a dual-duct network to mix air in the conditioned zone. Figure 5-17 shows a multizone unit that controls airflow by variable-volume dampers in the ducts.

Central Cooling Packages

Central cooling packages are used primarily in residential and light commercial building installations. Typical packages, such as the one shown in Figure 5-18, are of rectangular box construction. Supply-air and return connections are on the front of the unit with a condensing-air section on the back or side. Construction is such that return air is drawn through a filter and forced to pass through a fin-tubed evaporator coil by a centrifugal blower. Cool air is discharged from the supply duct for distribution throughout the system. In small units air circulation is provided by direct-drive blowers; larger units employ adjustable belt-driven blowers. The evaporator coil and compressor are usually located near the supply duct and are insulated from the rest of the unit to prevent heat gain and reduce sweating. The cooling capacity of central cooling packages ranges from 1.5 to 7.5 tons for residential use and up to 30 tons for commercial installations.

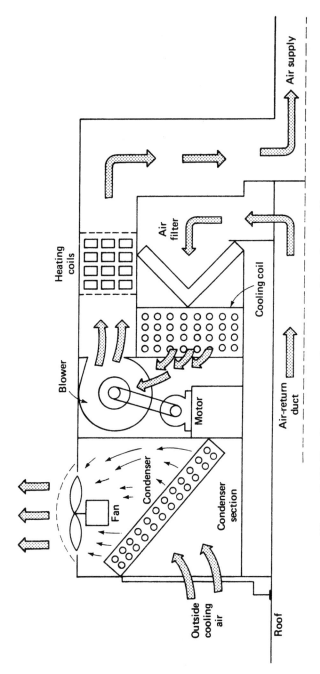

Figure 5-15. Rooftop-mounted year-round heating/cooling system.

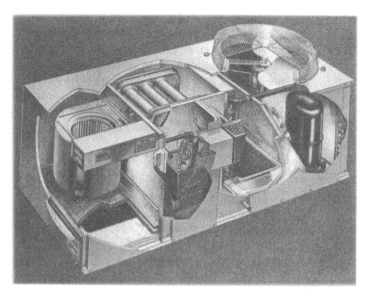

Figure 5-16. Single-package gas heating/cooling system. (*Courtesy of Lennox Industries, Inc.***)**

Figure 5-17. Multizone year-round conditioning system. (*Courtesy of Lennox Industries, Inc.***)**

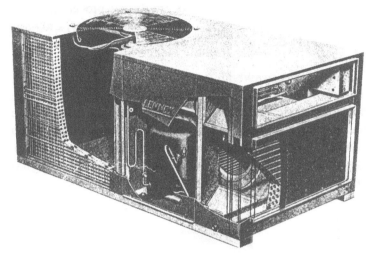

Figure 5-18. Central cooling package. (*Courtesy of Lennox Industries, Inc.*)

SPLIT-SYSTEMS

Split-systems are named because a portion of the cooling system is split, or mounted remotely from the other portion of the system. Typically split systems are selected when air cooling is desired or to be 'added-on' to a system that already has forced air, or some sort of air handling for an existing heating system. In such an installation, the cooling components (also referred to as the 'cool side') are added in to the existing air handling equipment. Components such as the evaporator and expansion valve are placed between the air filtration device and the heat source in the air handler. This allows all the air being returned to the system to flow over the evaporator. In times of heating need, the cooling system is turned off and the heat source (electric, gas, etc.) functions normally. However, in times when cooling is needed, the heat source is turned off, and the split cooling system is turned on, and the air passing over the evaporator is cooled. The other half of the split system, the exterior (also referred to as the 'hot side') contains the compressor, the condenser, and the condenser fan. This external unit, also called the condensing unit, can be mounted anywhere sufficient airflow is available, and can include the rooftop. Typically the only link between the two systems is pressure and return lines for the refrigerant, and some control wires. One special type of split-system the ductless split-system has recently gained in popularity. Sometimes nicknamed 'mini-splits', these systems have the same physical makeup as their full size counterparts with one important difference. The interior module does not require an air handler, the air movement across the evaporator is supplied by a fan built into the module.

Advantages of split-systems typically include the elimination of unsightly and noisy window air-conditioning units with one condensing unit, as well as allowing air cooling throughout the entire dwelling. The ductless split-system enjoys the added benefits of easier installation, zone control of temperatures, as well as increased interior design options because the modules can be mounted almost anywhere, including ceilings. Some disadvantages include cost, the rework necessary to fit the inside unit to the existing cabinet for the air handler used by the heating system, and the additional installation of a condensate drain.

AIR-CONDITIONING-SYSTEM COMPONENTS

The primary function of a summer air-conditioning system is to cool and dehumidify the inside air of a building or area. Cooling the air and dehumidifying it are both related to the heat-removal problem. To cool hot dry air, its sensible heat must be reduced. Sensible heat is primarily responsible for causing an increase in temperature. Our body senses the fact that air is hotter. Latent heat, on the other hand, does not cause an increase in the temperature reading of a thermometer. It is, however, responsible for heat that is present in the moisture vapor of the air.

Sensible heat is removed from the air of a building by causing it to flow from the warm air of a given space into a transfer medium that is carried away. In dehumidifying air, latent heat is removed from the moisture vapor of the air. Condensation from this vapor drains off into a drip pan, carrying latent heat with it. It must be kept in mind that a building is not being cooled or dehumidified—it is the air of the building.

The basic components of a summer air-conditioning system are responsible for pulling the inside air of a room into a heat-exchanging unit that will remove both sensible and latent heat. Sensible heat is carried away from the area by a gas that circulates through the heat exchanger. Cool air is returned to the room by a duct network with a blower assembly. The components of the system are essentially the same as those of a mechanical refrigeration system. This includes the evaporator coil, compressor, condenser, receiver tank, and control. A person working with this equipment must be able to recognize individual components and their function in order to evaluate system operation.

Compressors

The compressor of the mechanical refrigerator section of an air conditioner is considered to be the heart of the entire system. It is primarily responsible for the circulation and reuse of the refrigerant and is a mechanical means whereby work is done on the refrigerant. Through this action the temperature of the refrigerant is increased to a point where heat rejection is possible in the condenser.

Compressors for mechanical refrigeration equipment are often classified according to their method of operation. In this regard, the three most popular types of compressors are the reciprocating, rotary, and centrifugal. The type of system, its operational capacity, and its application has a great deal to do with the selection of a particular type of compressor.

Reciprocating compressor design is similar in many respects to the piston action of an internal combustion engine. Figure 5-19 shows a partial cutaway view of a three-cylinder hermetically sealed compressor. In this type of assembly, the motor and compressor are both sealed in the same housing. As a result of this construction, the motor is cooled by the refrigerant that flows through and around the housing during its operational cycle.

Reciprocating compressors employ one or more pistons that are driven by a common crankshaft. Rotation of the crankshaft causes the piston to make alternating suction and compression strokes inside a cylinder. When the piston is pulled down inside the cylinder, refrigerant vapor is drawn into the top of the cylinder. A reed-like flapper valve in the cylinder head opens during the stroke. This valve responds as a check valve that admits vapor to the top of the cylinder

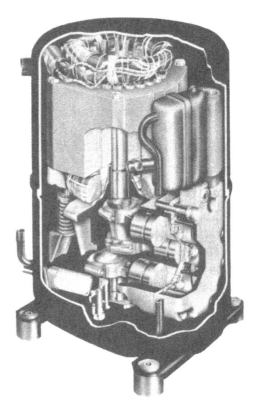

Figure 5-19. Three-cylinder sealed compressor. (Courtesy of Tecumseh Products Company)

only during the suction stroke. When the vapor pressure inside the cylinder eventually equals the refrigerant vapor pressure, the reed valve closes. This usually occurs when the piston reaches the bottom of the suction stroke. See the suction stroke of the compressor in Figure 5-18.

When the piston reaches the bottom of the suction stroke, it has drawn all the refrigerant vapor into the cylinder that it can handle. At this point of the operational sequence, the piston begins its compression stroke. The upward thrust of the piston into the cylinder causes the trapped vapor to be compressed into a smaller volume. Vapor cannot return to the evaporator because of the increased pressure buildup on the suction reed valve. As an end result, the vapor has

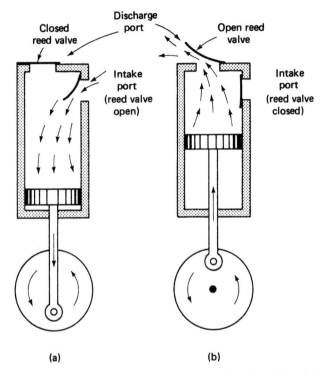

Figure 5-20. Reciprocating compressor operational cycles: (a) suction stroke; (b) compression stroke.

increased in pressure.

When the piston begins to reach the uppermost thrust of the compression stroke, a second reed valve is opened by the increased pressure. This valve is designed to release the compressed vapor at a specific pressure value. The discharged vapor then passes from the compressor into the condenser for the next step of the operational cycle. At this point the piston begins to move downward, starting the next suction stroke. See the compression stroke of Figure 5-20.

Centrifugal compressors are of the turbo family of machines, including fans, propellers, and turbines. Pumping force is created by the rotational speed of a turbine-like blade and the angular movement of the refrigerant flowing through a channel in the housing. Centrifugal compressors are not responsible for moving a specific amount of refrigerant during each rotational pass of the blade. As a result of this characteristic, they are classified as having a nonpositive displacement operational cycle.

Centrifugal compression is achieved by whirling the mass of a gaseous refrigerant at a high rate of speed and causing it to be thrown outward by centrifugal force. Channels around each blade have a diminishing area that compresses the refrigerant and causes it to have increased pressure. Each rotating blade is called a stage of compression, and typical units use five or more stages between the inlet and outlet ports. Figure 5-21 shows a cross-sectional view of a single compressor turbine blade. Refrigerant enters the center of the turbine and is forced to the outer periphery of the rotor by centrifugal force. It is then directed into a channel, where it is forced to the center of the next stage. A longitudinal section of a five-stage compressor is shown in Figure 5-22. Note the flow path of the refrigerant, indicated by the arrows.

Centrifugal compressors must operate at relatively high speeds to develop suitable operating pressures for air-conditioning systems. In general, this type of compressor is quite efficient and is well suited for large-capacity refrigeration units. They are widely used in systems having a capacity of 200 to 3000 tons.

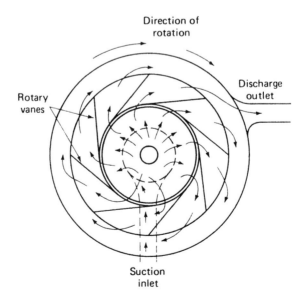

Figure 5-21. Front view of a centrifugal compressor blade.

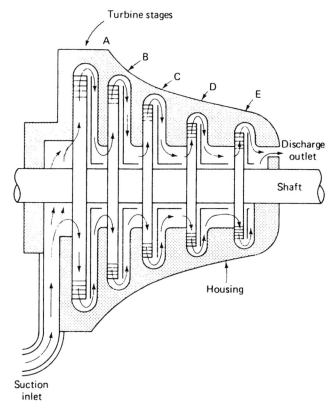

Figure 5-22. Longitudinal section of a five-stage compressor.

Rotary compressors employ a rather simple circular-motion principle to compress refrigerant vapor and increase system pressure. Compressors of this type employ a rolling piston that is driven by an eccentric roller shaft. As the piston rotates, vapor is drawn into the space ahead of a spring-loaded wiper blade, as shown in Figure 5-23. Compression takes place when the roller passes into a continually smaller space until it is forced out of the discharge port. After the piston passes the wiper blade, the compression cycle begins again by taking in a new supply of refrigerant vapor.

Rotary compressors are classified as positive displacement units because they move a specific amount of refrigerant vapor with each pass of the roller piston. This type of compressor is usually hermetically sealed in a shell housing with a direct-drive motor. The entire assembly

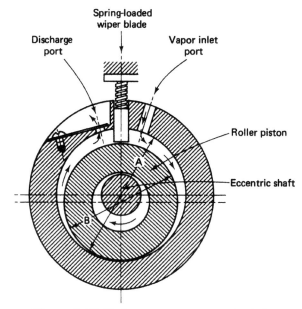

Figure 5-23. Rotary-compressor principle.

is cooled and lubricated by the refrigerant vapor. Rotary compressors are commonly used in high-capacity air-conditioning systems.

Scroll compressors rather use spiral-shaped devices to compress the refrigerant rather than pistons or vanes. In the compressor, one spiral, or scroll, is fixed and the other scroll is allowed to rotate. As the refrigerant is allowed to pass into the compressor, it is trapped in smaller and smaller areas toward the center of the spiral-shaped scroll. The compressed refrigerant is then released from the center of the scroll. The primary advantages of scroll compressors over other conventional compressors are reduced noise, increased dependability, and increased efficiency. The use of scroll compressors is one of the methodologies by which air conditioning manufacturers are utilizing to increase the efficiency in order to meet government standards.

Condensers

The condenser of an air-conditioning system is primarily responsible for removing heat from the refrigerant after it leaves the compressor. Basically, this part of the system is classified as a heat-exchanging unit. Heat picked up by refrigerant passing through the evaporator causes it to vaporize. Vapor heat plus heat caused by the compressor must both be dissipated by the condenser. When the temperature of the refrigerant passing through the condenser is lowered to its saturation point, it causes the vapor to change into a liquid. The receiver then accepts this liquid at the outlet side of the condenser and stores it for reuse during the next cycle of operation.

The cooling effect of a condenser may be achieved by air, water, or evaporation. Residential air conditioners and many low-capacity units employ air as the primary cooling medium. Condensers of this type must be located in a position where they receive an abundance of outside air.

Water is also used as a condenser medium in many systems. This type of unit uses water to carry heat away from the refrigerant. Water usually circulates through the outer shell of the condenser, with refrigerant passing through internal pipes or tubes. Cooling by this method is generally found in systems with moderate to high cooling capabilities. Evaporation is the third approach to condenser cooling. Both water and air are used in medium-tonnage systems where water problems are often a factor. The cooling medium of the condenser represents one of the most distinguishing features of an air-conditioning system.

Air-cooled condensers are widely used in small commercial building units and residential air-conditioning systems. Typical construction of this condenser includes a metal tube coil with external fins attached for heat dissipation. Circulating air forced to pass through the structure carries heat away from its surface. Figure 5-24 shows the operating principle of an air-cooled condenser.

Figure 5-25 shows the outside unit of an air conditioner using an air-cooled condenser. This particular unit houses the condenser, compressor, and fan-motor assembly. In this case, condenser coils are located on each side of the housing. Air is drawn through the condenser fins by the fan and forced out of the housing through the grille at the top. Operation is dependent upon an ample supply of "cool" air to dissipate heat. Even when the air is in excess of 100°F (38°C), it is cooler than the heated refrigerant passing through the condenser coil.

Air-cooled condensers are easy to install, inexpensive to maintain, require no water for operation, and present no danger of freezing in cold weather. Efficient operation is based upon a sufficient quantity

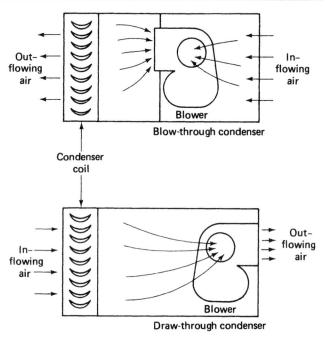

Figure 5-24. Airflow through an air-cooled condenser.

Figure 5-25. Air-cooled condenser assembly. (Courtesy of Lennox Industries, Inc.)

of fresh air being pulled through the condenser. In many installations this may create a noise problem with the fan or blower assembly. By locating the condenser unit in a remote outside area or on the roof, the noise problem can be minimized.

Water-cooled condensers are normally used in air-conditioning systems where an adequate supply of low-cost water is available. This type of condenser generally provides a rather low condensing pressure, which permits better control of the refrigerant. In most applications water is usually much cooler than normal daytime ambient-air temperatures. Water is also a more efficient medium than air for transferring heat. Representative water-cooled condensers are generally quite small and compact compared with condensers using another medium.

Water-cooled condensers are classified today according to their physical construction. The housing and internal construction of the condenser are both included in this classification. Shell-and-tube, shell-and-coil, and tube-in-tube are three of the most popular types of condenser construction. Small- to medium-capacity units employ the shell-and-tube type.

Shell-and-tube condensers consist of a cylindrical steel shell containing several interconnected copper tubes running parallel through the center of the shell. Water is forced into the tubes at the inlet and is removed at the outlet port. Hot refrigerant vapor is normally applied to the condenser at the top of the shell. After cooling, liquid refrigerant begins to flow toward the bottom of the shell. This type of condenser therefore has a dual role. Its primary function is as a condenser and it responds by cooling the refrigerant. In addition to this, it serves as a receiver tank for liquid refrigerant storage. Figure 5-26 shows a simplification of the shell-and-tube type of construction.

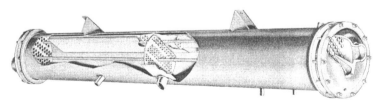

Figure 5-26. Internal view of a shell-and-tube condenser. (*Courtesy of York International Corp.***)**

Metal plates bolted to the ends of the shell-and-tube condenser provide easy access for cleaning of the tubes. This must be done periodically to remove mineral deposits that accumulate on the inside of the tubes. Excess tube scaling tends to restrict water flow, which reduces the transfer of heat.

Water control of the shell-and-tube unit is based upon the travel length of the flow tube and the number of passes it makes through the shell. When water enters at one end and flows once through the tube to an outlet on the other end, it is classified as a one-pass condenser. Units having the water inlet and outlet on the same end plate must pass two or some other even number of times through the shell. Two-pass condensers are widely used today in many low-capacity cooling systems.

Shell-and-coil condenser construction is unique in design compared with shell-and-tube units. Instead of straight lengths of tubing, a coil of tubing is formed and placed in a rather small shell housing. Condensers of this type are usually quite compact in size and have a rather high level of heat-transfer ability. Construction of this type also permits the unit to be used as a combination condenser and receiver storage tank. Figure 5-27 shows a simplification of the shell-and-coil

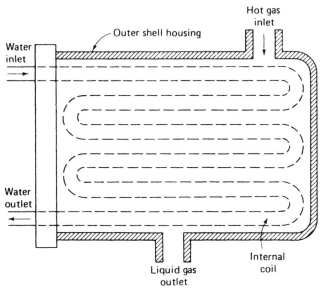

Figure 5-27. Simplified shell-and-coil condenser.

condenser.

Shell-and-coil condensers are rather widely used in low-capacity cooling applications which can be assured of a reasonably clean supply of water. Cleaning of the flow coil is somewhat more difficult to achieve than with tube construction. Removal of mineral deposits can be accomplished only by a flushing procedure that uses chemical cleaners. Condensers of this type must be cleaned on a regular basis to assure efficient operation.

Tube-in-tube condensers are widely used in water-cooled air-conditioning equipment. Condensers of this type are constructed from a series of small-diameter tubes placed inside larger-diameter tubes. As shown in Figure 5-28, refrigerant vapor is applied to the larger-diameter tube with cool water flowing through the smaller inside tube. Heat transfer takes place between the cool inside tubes and air circulating over the larger-diameter outside tubes. In a sense, condensers of this type respond to both air and water cooling action.

Installation of a tube-in-tube condenser has cool water entering the bottom of the assembly and being removed at the top. The refrigerant is applied in a reverse direction. It enters at the top of the unit and is removed in liquid form at the bottom. Construction of this type is designed to provide a rather high level of heat-transfer efficiency. The coldest water is capable of removing a great deal of heat from the

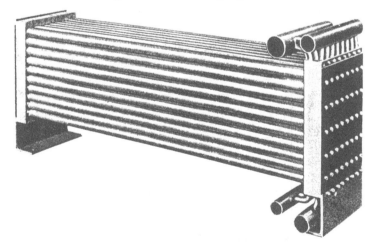

Figure 5-28. Tube-in-tube water-cooled condenser. (Courtesy of Halstead and Mitchell Co.)

refrigerant after it has been changed into a liquid state. This action is often called *subcooling*. Warmed water circulating toward the top of the unit still has the capability of removing high-temperature vapor heat. The operational efficiency of this type of construction makes it particularly well suited for condensers of high-capacity cooling systems.

Evaporative condensers are frequently used in applications where satisfactory condensing action is difficult to achieve due to high ambient-air temperatures and where the supply of water is not adequate for heavy usage. Condensers of this type combine the use of both air and water in their operation. Figure 5-29 shows a functional diagram of an evaporative condenser.

An evaporative condenser consists of an air-cooled condenser over which a spray of water is allowed to flow. Air is drawn into the assembly from the bottom, flows upward across the condenser coil, through the spray, and exits through the top of the unit. Heat from refrigerant vapor passing through the condenser coil is transferred to the water that wets its outer surface. Heat is then transferred to the circulating air as it evaporates. The source of air is either blown or drawn through the water spray by circulating fan. The eliminator plates remove water from the exhausting air so that mineral deposits or scale will not build up on the fan blower assembly. Figure 5-30 shows a typical evaporative condenser assembly looking into the air handler.

Liquid Receivers

Many water-cooled condensers serve as a combination condenser and liquid refrigerant receiver. In this type of component, after the refrigerant is changed into a liquid, it is stored in an available space at the bottom of the condenser. Combination units are generally designed so that they can accumulate a complete charge of system refrigerant. If the amount of refrigerant to be stored requires more space than that provided by the condenser, a separate receiver tank must be employed. The amount of refrigerant required for system operation is determined primarily by the cooling capacity of the system. If more than 8 lb. (3.63 kg) of refrigerant appears at the output of the condenser, it should be stored in an independent liquid receiver tank as required.

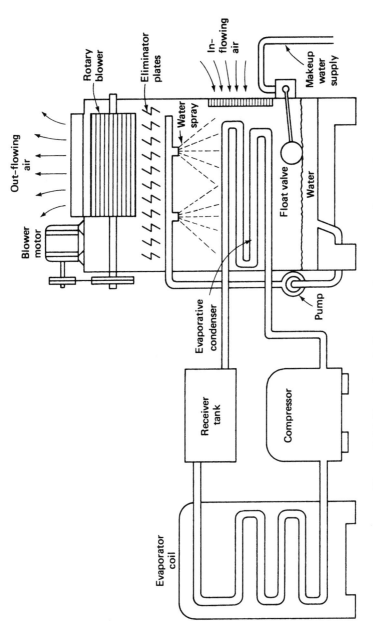

Figure 5-29. Functional diagram of an evaporative condenser.

Figure 5-30. Air-handler output of an evaporative condenser. (*Courtesy of Halstead and Mitchell Co.*)

Receiver tanks are normally installed in the refrigerant discharge line of the condenser. This line connects to the top of the receiver and extends downward into the tank to the liquid-level point. In operation, it is important that the level of the refrigerant never fall below the bottom of the extension tube. It is also important that the tank not receive too much refrigerant. This would interfere with the normal gas storage space above the liquid and result in excessive discharge pressure.

In large-capacity air-conditioning systems, the receiver tank must incorporate a safety protection device to prevent possible damage due to excessive pressure. These devices are normally in the form of a fusible metal plug installed in the end of the tank. Exposure of the tank to excessive heat, such as a fire, would cause the fusible plug to melt. Receiver pressure would be released harmlessly without the danger of an explosion. The melting point of a fusible plug is in the range 165 to 212°F (74 to 100°C).

Evaporators

The evaporator of an air-conditioning system is primarily responsible for removing heat from a specific area or room and to produce cooling. When a refrigerant enters the internal flow path of the evaporator coil, it absorbs heat from air passing over its outside surface. This action causes the refrigerant to boil and immediately turn into a vapor. The suction line of the compressor then pulls this vapor from the evaporator, where it is increased in pressure and temperature before being applied to the condenser. Through this action, heat is transferred from the circulating air, which in turn causes cool air to flow back into the area being conditioned. Evaporator coils are an essential part of both mechanical and absorption refrigeration systems.

Evaporators differ a great deal today in physical construction, shape, and size. For air-conditioning applications, evaporators are either of the direct expansion type or the water-chiller type.

Direct expansion units are used for cooling air that comes in direct contact with its outside surface. *Water chillers* are designed to cool fresh water that ultimately circulates through pipes or tubes. Both evaporator types are designed for efficient heat transfer and for maximum pressure drops through the coil.

Direct expansion evaporators are constructed of copper tubes that are held in place by a large number of thin metal fins. The fins are designed to extend the surface of the coil so that there will be a more effective transfer of heat from air to the coil. The coil tube provides a continuous path for refrigerant to pass between the flow control device and the suction line of the compressor. The flow of refrigerant through the coil causes it to absorb heat and be changed into a vapor. Figure 5-31 shows an example of a direct expansion cooling coil. Note the

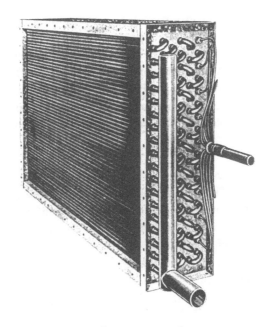

Figure 5-31. Direct expansion evaporator coil. (*Courtesy of York International Corp.*)

close spacing of the metal fins and the crimped construction that increases the surface metal of each fin.

Direct expansion evaporator coils are divided into two distinct groups according to the method used to handle refrigerant. Dry-type coils are designed so that they do not collect refrigerant in a pool for recirculation. Flooded coils by comparison employ a separation chamber. As liquid refrigerant flows through the evaporator, it absorbs heat and changes into a vapor. Upon leaving the evaporator, any liquid refrigerant that is still present is separated from the vapor and accumulates in a chamber on the outlet side of the evaporator. The separated liquid is recycled by returning it to the evaporator input. Flooded evaporators generally provide better heat transfer by assuring that the inside surface of the coil is always in direct contact with the liquid refrigerant for all conditions of operation.

Water-chiller evaporators are used primarily in large central-station air-conditioning systems. In this type of system, chilled water serves as the heat-transfer medium. Water is chilled to approximately 45°F (7°C) and then circulated to individual rooms or space cooling coils, where it absorbs heat. Upon returning to the chiller, water heat is transferred to the refrigerant, which causes it to vaporize. The evaporator function of this system is accomplished by the cooler unit of Figure 5-32. Note that this cooler is of the shell-and-tube type of construction.

High-capacity water-chiller systems usually employ the flooded type of cooler in their construction. In this assembly, water being chilled flows on the inside of tubes while the outside surface is covered with liquid refrigerant. Boiling occurs when tube heat is transferred to the refrigerant. Vapor formation around the outside area of each tube rises to the top of the cooler, where it is collected by the suction line of the compressor. See the cutaway view in Figure 5-33. This particular type of system is available in sizes up to 630 tons.

Water-chiller evaporator action for small- to medium-capacity units is generally achieved by dry-expansion chillers. In this type of system, shell-and-tube construction is commonly used. Refrigerant circulates through small internal tubes and water circulates through the surrounding outside shell. See the cutaway view in Figure 5-34.

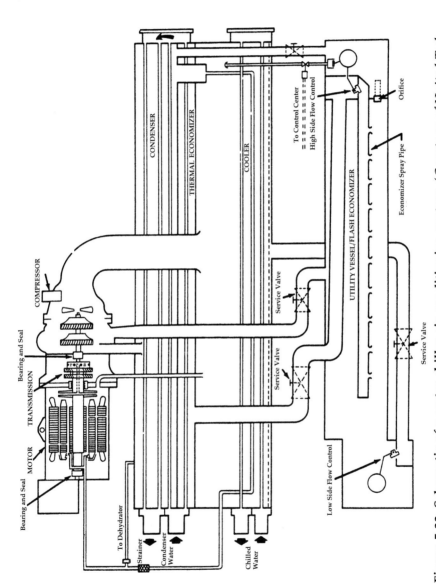

Figure 5-32. Schematic of a water-chiller air-conditioning system. (Courtesy of United Technologies Carrier Corporation.)

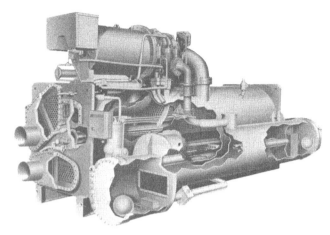

Figure 5-33. Cutaway view of a water-chiller air-conditioning system. (*Courtesy of United Technologies Carrier Corporation.*)

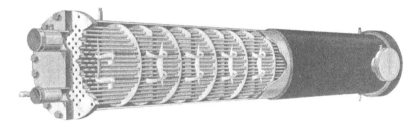

Figure 5-34. Cutaway view of a dry-expansion water chiller. (*Courtesy of York International Corp.***)**

EFFICIENCIES IN AIR CONDITIONING

Once the general concepts of air conditioning are understood, it is important to be able to select the type of system that is most efficient for the application. This section discusses some general methodologies for increasing efficiency as well as determining efficiency of systems.

Aside from a proper thermostat setting, the best place to increase the efficiency of an air conditioning unit is to perform proper maintenance. Significant amounts of energy are needlessly expended because of poorly maintained systems. Lubrication and making sure all system filters are clear, cleaned, or replaced regularly can cut costs in energy consumption and repair bills. Too, units should be regularly checked for dirty coils—particularly in exterior units—which inhibit air flow, and decrease efficiency. Also, leaks in ducts or refrigerant can cause air conditioning systems to be overworked, and possibly environmental harm. Clearly the least expensive method to improve the efficiency of an existing system is to provide proper scheduled maintenance.

As previously discussed in this book, one measure of efficiency is the EER, or the energy efficiency rating. The EER is determined by taking the Btu rating of an unit divided by the power it consumes (in watts). Higher EERs indicate higher efficiency. However, a new measure of efficiency has been adopted by the government and equipment manufacturers, the SEER, or seasonal energy efficiency ratio. This rating, mandated by the Department of Energy, is to be assigned to every unit sold in the U.S. Units produced after 1992 had to meet a minimum SEER rating of 10, and as of January 2006, units must meet a minimum rating of 13 SEER. Because units capable of

Figure 5-35. A modern chiller

higher efficiencies are typically more expensive to produce, they have higher cost. On the other hand, a higher SEER rating means higher energy cost savings. An air conditioning professional can calculate the proper SEER rating for any application. In addition to SEER rating, consumers may look for equipment that is energy star compliant. For example, energy star qualified central air conditioners are about 20% more efficient than standard efficiency products.

REFRIGERANTS

The refrigerant is the medium by which heat is transferred in air conditioning systems. This classification of materials has the unique ability to give up heat when changing from a liquid to a gaseous or vapor state. Typically these were from a chlorofluorocarbon (CFC) family of compounds. Once it was discovered that compounds containing chlorine are dispersed into the atmosphere, and have the ability to damage the layer of ozone in the stratosphere, the government began regulating the production and sale of refrigerants. This destruction is significant because one atom of chlorine can break apart 100,000 ozone molecules. Early refrigerants commonly contained CFCs, but are being phased out and replaced by hydrofluorocarbon (HFC) refrigerants, which have no chlorine. Typically, refrigerants are identified by a code number such as R12 or R134a. These refrigerants are coded in this manner:

- Ones digit- number of fluorine atoms per molecule

- Tens digit- number of hydrogen atoms per molecule plus one

- Hundreds digit- number of carbon atoms per molecule, minus one

- Suffix of a,b,c- indicates unbalanced isomer molecule

For example, R134a has 4 fluorine atoms, 2 hydrogen atoms, 2 carbon atoms, with an isomer molecule of tetrafluoroethane

For many years R12 was used exclusively in automobile and portable air conditioning units and R22 in residential and light commercial applications. As the connection between the ozone depletion and use of CFCs was made, refrigerants with chlorine such as R12 were phased out, and safer refrigerants such as R134a were instituted. Soon R22 will stopped being produced in 2015, and will be phased-out of residential systems to be replaced by R410A. Other refrigerants such as R502 were required by the EPA to halt production, and systems were redesigned to accept 'zero ozone depletion' refrigerants such as R404A.

INDOOR AIR QUALITY

As heating and air conditioning systems improve in quality and efficiency, another consumer trend has been identified, indoor air quality or IAQ. At the time of this manuscript, this market of accessories to heating and air conditioning systems is on the rise, and will likely reach its peak in about the year 2010. Concerns that govern indoor air quality involve filtering or eliminating mold and allergens, CO_2 content, and humidity. While it is true that many heating and air conditioning systems now require air filtration, it is not of the level that will be asked for by consumers in the coming years. Each of the technologies currently used to address IAQ will be discussed.

Filtration

Currently filtration is used to remove dust from air handling systems. However, these filters are very coarse by nature and do not block the smaller particles related to mold and allergens. Other types of filtration such as HEPA (high efficiency particulate absorbing) that are capable of removing over 99.9% of the particulate matter from the air are added to heating and air conditioning systems to improve IAQ. In addition to a HEPA filter, other air-handling-equipment filters can be fitted which have a minimum efficiency reporting value or MERV rating. MERV ratings are from 1 to 16 and reflect the filter's ability to remove particles. A low MERV number indicates a fairly coarse filter, and a filter with a MERV rating of 14 might be used in a hospital where airborne bacteria are a concern. Because a MERV rating is an industry standard, it is an effective method to use when evaluating filters from different manufacturers. In addition to filtration, UV (ultraviolet) lighting can be used to inhibit airborne biological hazards. UV systems can be installed in plenums, air handlers or ductwork for disinfection. By passing harmful germs past the UV light source, the DNA of the germ is destroyed. This technology has been proven effective in the treatment of viruses, molds, spores, fungi, and bacteria. As with many IAQ technologies, filtration and UV lamps have costs that increase as IAQ increases. In filtration, filters with higher MERV ratings cost more and restrict airflow, thereby decreasing efficiency. In UV light treatment, in addition to the initial cost of installation, the UV lamps have a relatively short life, sometimes as brief as one year.

Humidity

An acceptable level of humidity in air that has been heated or air conditioned can have an effect on how warm or cool the air feels to humans. In winter heating, a heating system operating at 68 degrees Fahrenheit with 60% relative humidity provides the same level of comfort as a system set at 72 degrees Fahrenheit at 30% humidity. Therefore, adding moisture to the air can result in creature comforts with lower energy expenditures. In cooling season, the opposite is true. An air-conditioning system operating at 78 degrees Fahrenheit at 30% humidity provides the same level of comfort as a system operating at 74 degrees Fahrenheit and 70% humidity. In summertime, the lower level of humidity will increase comfort and lower bills. Humidification is primarily achieved by four primary methodologies. The first and perhaps most common is the flow-thru humidifier. In this method, a media capable of aiding in the evaporation of water is placed in the direct path of air flow. Through some sort of continuous wetting process, the media is wetted, and evaporation improves the humidity. Another method, spray humidification simply directly sprays water into the flow of air to improve evaporation. Drum-type humidifiers allow air to be passed through a drum-shaped object that is slowly rotated through a bath of water. The evaporation of water from the slowly rotating drum increases the humidity in the air that passes over it. Perhaps the best and most costly method of humidification is steam. In a steam humidification system, a heating element provides a source of heat energy that boils water in which it is immersed, or that passes over it, and the steam discharged by this system is allowed to enter the flow of air. Dehumidification can be obtained by mechanical devices called dehumidifiers that use air conditioning technologies to desiccants, materials to absorb the water or humidity from the air.

Carbon Dioxide

Another indicator of air quality is the level carbon dioxide, or CO2 in a room. Because humans take in oxygen and exhale CO2, the level in a closed chamber or room can deteriorate over time if fresh air is not brought into the system. To improve this, sensors that can detect carbon dioxide are placed in a room. When a certain threshold of CO2 is reached, ventilation allowing outside air to be brought in are opened. The exact amounts of CO_2 allowable and formula for calculating the rate of fresh air needed can be found in the literature concerning ventilation rate procedures, particularly ASHRAE Standard 62, Ventilation for Acceptable Indoor Air Quality. It should be noted that CO_2 levels are not considered an IAQ indicator by ASHRAE Standard 62, but is often referred to when researching air quality considerations.

COMMERCIAL AIR HANDLING

To better illustrate how some of the systems mentioned in earlier in this text, consider the following example. The following figures illustrate a complete air handling and energy recovery system for a commercial installation. Much like all systems, air is introduced into the unit, filtered, heated or cooled and returned to the supply air to the building after being properly filtered and conditioned for proper temperature.

The air comes into the unit in Figure 5-36. The incoming air is made up from two sources, fresh air from outside enters from the dampened intakes on the right, and return air from the building from the vents above. The bladed objects on the left are not fans, but static blades capable of mixing the return and outside air with a minimal drop in pressure. For scale, note the stepladder and the door on the far end for servicing personnel.

After the air is blended, it is passed through an initial filter system. This removes any contaminants that may have entered from both air sources, and aids in keeping the air handling equipment clean,

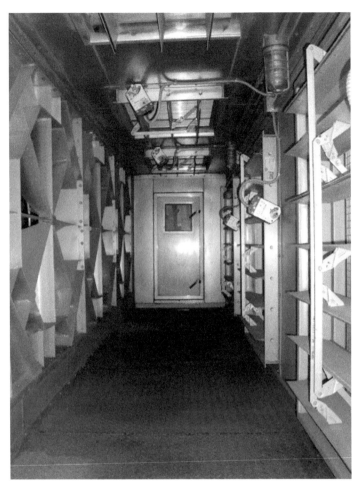

Figure 5-36. Mixing of ourside and return air.

thereby increasing life and lowering maintenance costs. This transition is shown in Figure 5-37, the pleated wall of filters is on the right.

Once filtered, the air moves over coils from an energy recovery unit. These coils are connected to another unit located where heating or cooling air is exhausted from the building. To operate, the exhaust air passes over glycol filled coils, thereby extracting heat in the heating season or cool in the cooling season. As the energy is recovered into the glycol, it is pumped back to the energy recovery section of the air handler and aiding in heating or cooling in a manner to pretreat the incoming air. Figure 5-38 shows this heat exchange on the right.

After pretreatment, by the energy recovery unit, the air passes over the heat exchanger coils. If it is heating season, these coils will be filled with a heated fluid supplied by a heat exchanger with the heat supplied by a boiler. As with most heat exchangers, the heat is extracted by the coils through the fins, heating the air as it passes. In the cooling season this bank of heat exchanging coils is not used. The heat exchanger on the left of Figure 5-38 are the heating coils.

After the heating section, air passes over the cooling section. Here treated water from the chiller is passed through the coils of the heat exchanger, thereby cooling the air that passes over them. Similar to the heating coils, this section is only used during the cooling season. The transition between heating and cooling coils is shown in Figure 5-39.

As described earlier, a blower or fan is typically used to move air through the coils and filters in an air handler. In this commercial installation, a large bank of fans draws air through this air handler. A bank of eight (four fans on two levels) large motors drives fans that

Figure 5-37. Pre-filtering the mixed air.

Figure 5-38. Energy recovery coils on the left; preheat coils on the right.

Figure 5-39. Transition betwen preheat coils on left to cooling coils on the right.

pull air through the air handler and ultimately supply force to deliver air throughout the building. Some of these fans are shown in Figure 5-40, note the large vibration dampening mounts for each blower motor assembly.

Before the air exits the air handler, it is filtered. This system uses twelve inch MERV 15 filterers that are 12 inches thick. Figure 5-41 (a) and (b) show these filters removed from the system. From this location the air is distributed throughout the building.

ENERGY RECOVERY SYSTEMS

Because commercial systems require a frequent and constant exchange of indoor air with outdoor air, air that has been conditioned for heating or cooling season (and the money required to heat or cool this air) is lost at discharge. Vented air without an energy recovery system operates similar to a bathroom vent fan in a residential installation. Many types of energy recovery systems have been developed to recover this heating or cooling energy from the air prior to discharge. Instead of direct discharge of the temperature conditioned air, energy recovery systems use a heat exchanger to extract heat in the

Figure 5-40. Fans and fan motors

Figure 5-41a and b: Commercial MERV 15 filters.

heating system or cool in the cooling season and use this extracted energy to pretreat air in the traditional air handler.

Using energy recovery systems have two advantages. The obvious advantage of the energy recovery system is the extraction of the energy that would have normally been discharged. Fewer losses mean lower expenses in heating or cooling the air. A second advantage is the allowance for using reduced capacity equipment, lowering both operating and capital costs.

The large-scale energy recovery system presented in Figure 5-42 shows inside the heat exchanger and where the air about to be exhausted is passed over the energy recovery coils on the right. These coils are filled with a fluid (typically a mix of water and glycol) that accepts the energy until it is pumped back to an air handler for pretreatment of incoming air. The final discharge is through a roof mounted exhaust fan unit shown in Figure 5-43. Controls for this system are computerized and the motors that run the blowers with variable frequency drives. A display of the VFD is shown in Figure 5-44.

Figure 5-42. Inside the energy recovery unit.

Figure 5-43. Exhaust fans, roof mounted.

Figure 5-44. The display for a variable frequency drive (VFD) unit.

ENERGY CONSERVATION CHECKLIST FOR AIR-CONDITIONING SYSTEMS/SAVING NATURAL RESOURCES

Table 5-1 presents a checklist for air-conditioning systems.

Table 5-1. Checklist for Air-Conditioning Systems/Saving Natural Resources

Items to check	Corrective action
General Operational Procedures	
Evaluate the room space or area cooling schedule for those areas used infrequently or for only short periods of time.	Check the individual room and space utilization schedule. Shut off or reduce cool-air flow to rooms not being used.
Check the cooling temperature thermostat setting of offices, occupied work spaces, and residential areas.	Inspect thermostat setting and adjust to 78 to 80° when occupied. Special consideration should be given to rooms housing computers, cooking facilities, and process equipment.
Examine the cooling schedule of the building during evening hours and weekend operational periods when it is unoccupied.	Adjust the cooling schedule to higher levels for weekends and during evening-hour operation. Begin precooling operations so that the building is 80°F when the occupants arrive.

Table 5-1. Checklist for Air-Conditioning Systems /Saving Natural Resources (*Continued*)

Items to check	Corrective action
General Operational Procedures	
Inspect the cooling of lobbies, corridors, and vestibules.	When appropriate, reduce the flow of cool air to lobbies and corridors by closing supply ducts and registers.
Check the operational time of small unit or room air conditioners.	Set up an operational schedule for the use of small unit or room air conditioners.
Check the running speed of system ventilating fans and cooling units.	In mild weather it may be possible to lower the cooling effect by running room cooling fans at lower speeds.
Check the generation of unnecessary heat from lighting, cooking equipment, and machinery.	Turn off unnecessary lighting and reduce equipment operational time when at all possible.
Equipment Operational Procedures	
Examine the air-conditioning equipment operational instruction manual and nameplate data.	Establish a routine equipment maintenance checklist. Check for: 1. Correct motor voltage and current values. 2. Motor contactor connections and contact points. 3. Unbalanced voltage values on the three-phase power line.

Table 5-1. Checklist for Air-Conditioning Systems /Saving Natural Resources (*Continued*)

Items to check	Corrective action
	4. Improper equipment-grounding procedures.
	5. Belt alignment and condition.
	6. Motor, compressor, and rotating machinery bearing lubrication.
Check time clocks that automatically control operational periods and equipment operation sequencing.	Adjust and correct all time clocks and other control equipment for correct time of day and nighttime operation. Protect equipment from unauthorized adjustments.
Evaluate the actual temperature of rooms, areas, and space controlled by a thermostat setting.	Measure the temperature of spaces controlled by a thermostat and compare with set-point values. If necessary: 1. Recalibrate the thermostat. 2. Evaluate airflow or cool water path.
Evaluate the normal operational time of air-conditioning equipment. Does it run continuously for short operational cycles, or operate according to load demands?	Consult with the building engineer or equipment operation manual to become familiar with normal operational procedures. Report any unusual equipment operational conditions.

Table 5-1. Checklist for Air-Conditioning Systems /Saving Natural Resources (*Continued*)

Items to check	Corrective action
Equipment Operational Procedures	
Inspect and evaluate the flow of cool air through the ventilation duct network.	Clean and replace air filters at least once a month during the cooling season.
Inspect and evaluate the flow of cool air through the ventilation duct network.	Relocate furniture or equipment that might hinder the free flow of air through evaporator coils. Clean dust and dirt accumulation from evaporator and condenser coils or water-tower flow nozzles.
Inspect cooling lines and duct insulation between the compressor and evaporator.	Replace or repair damaged insulation to improve operational effectiveness.
In chilled-water systems, trace the flow path of chilled water and inspect for water leaks or condensation buildup.	Repair leaks or report the problem for maintenance. Clean dust and dirt accumulation from condensate drain or line.

Chapter 6

Lighting Systems/Improved Efficiency

INTRODUCTION

Lighting systems represent an important piece in the overall building puzzle. This system represents approximately 10% of the total electric energy load in the country. More than 20% of the electricity used in the home goes into lighting and in a commercial building this represents much more. While electric light bulbs are not expensive, they do contribute to your total energy costs.

Lighting systems convert electrical energy into light energy. Lighting systems are designed to create a comfortable and safe environment in buildings. There are several types of lighting systems in use today; among them are the incandescent, fluorescent, mercury-vapor, metal-halide, and several others. The planning involved to obtain proper lighting design in buildings is complex and may involve several different types of lighting. It is important that persons interested in energy management be familiar with the various types of lighting systems and their comparative energy use.

CHARACTERISTICS OF LIGHT

When dealing with light, there are basic terms that should be understood. The unit of light intensity is a standard light source called a candela or candlepower. The intensity of light is expressed in one of these units. The amount of light falling on a surface, all points of which are a unit distance from a uniform light source of 1 candela, is 1 lumen. The illumination of a surface is the number of lumens falling on it per unit area. The unit of illumination is the lux (lumens pe99square meter) or footcandle (lumens per square foot).

Reflected light, the light that is seen, is measured in candelas per square meter of surface (or footlamberts). The reflection factor is the percentage of light reflected from a surface (expressed as a decimal). The light reflected from a surface is equal to the illumination of the surface times the reflection factor. The brightness of a surface that reflects 1 lumen per square foot is 1 footlambert.

Characteristics of Daylight

The most readily available source of lighting is in the form of daylight. While a somewhat variable supply of light, it is often preferred because humans appear to prefer the connection to 'real' or 'natural' lighting. However, often this preferred light is variable due to the climate, time of day, the season, the orientation of the building, and the position of the user in relation to the earth's axis. Because of these characteristics, natural light may contain too many variables for use in some indoor lighting applications. It is also important to note that with daylight comes heating, which may not be a desirable effect in the summer months. Through proper types of glazing window treatments, this heating effect can be reduced.

In addition to visible light, daylight contains ultra-violet and infrared light that are outside the visible light spectrum. However, it shares this characteristic with many electric lamps used with artificial lighting, they also transmit light outside the visible range. Therefore, when examining the characteristics of daylight, not only are the specifications associated with light intensity such as lumens or candles, but the energy available as heat must also be considered. Typically this is measured the same as other units of heat energy, the British thermal unit, or Btu. It is estimated that one square foot of bright sunlight contains approximately 240 Btu of energy per hour.

TYPES OF LIGHTING

When planning a structure, several important decisions must be made in regard to lighting decisions. Generally, lighting is selected by the type of space, and the type of task or job performed in the space. Assembly areas of factories have different lighting needs than a television viewing area or parking lot. While three primary methods of lighting (fluorescent, incandescent, and halogen) have dominated the market, some changes in the technology of lighting may bring about a fourth method, the LED.

Daylight

Perhaps the oldest form of lighting is that supplied to us by the sun, titled ambient light or daylight. Beyond understanding the physical characteristics of daylight, it is important to understand the variability of the resource as well as the difficulty in controlling daylight as easily as constant as the controls available for artificial light. In addition to adding heat, natural light has the ability to produce a glare that can reduce usability of reading paperwork or a computer monitor. Both the variability and glare can make daylight less than ideal for many commercial and business applications. Such problems could lead to errors or user fatigue. Also, not all work done inside a structure is during the hours of daylight, and this resource is not available following sunset until sunrise the next day. Other concerns involve the design of the building such as bringing the daylight to interior rooms, or on floors below ground level.

On the plus side, the resource is essentially free. There is no electricity necessary for the production of the light, therefore monies in energy costs are saved. Methodologies to collect (such as trackers or concentrators), control (drapes, photosensors), distribute (fiber optics or light pipes), and deliver (diffusers) the light to the end user do have costs associated with them, but the actual resource, the daylight is essentially delivered free of charge to the site each day.

In order to bring daylight into the building, many strategies may be considered. Larger windows, changes in architectural design, adding ambient light collection structures such as atriums, skylights, or domes. In addition to these structural features, some of the problems earlier described such as heat and glare can be addressed by using glazing, coatings, or films to the window or point of light entry. A new plastics technology that holds great promise is a "smart window" that can change in response to changes in light intensity, heat, or an applied external voltage. Conventional methodologies for controlling the daylight involve the use of drapes or blinds.

One of the best solutions at this point involves hybrid type systems that take advantage of microprocessor based controls in which various sensors are placed in the room, and the computer maintains a unvarying level of light by maintaining a balance of natural and ambient levels by supplementing with artificial lighting.

INCANDESCENT LIGHTING

Incandescent lighting is a common type of lighting and its construction is shown in Figure 6-1a. Some common bases and bulb designations are shown in Figure 6-1b and c. Incandescent lamps are simple to install and maintain. The initial cost is low, but incandescent lamps have low efficiency and a short life span.

Incandescent lamps usually have thin tungsten filaments such as those shown in Figure 6-2. The filament is connected through the lamp base to an electrical power source. When an electrical current passes through the filament, the temperature of the filament rises to between 3000 and 5000°F (1649 and 2760°C). At this temperature range, the tungsten produces a high-intensity white light. When an

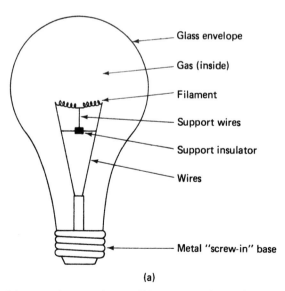

(a)

Figure 6-1. (a) Incandescent lamp; (b) common lamp bases; (c) common bulb designations. (b, c, *courtesy of Philips Lighting Corp.*)

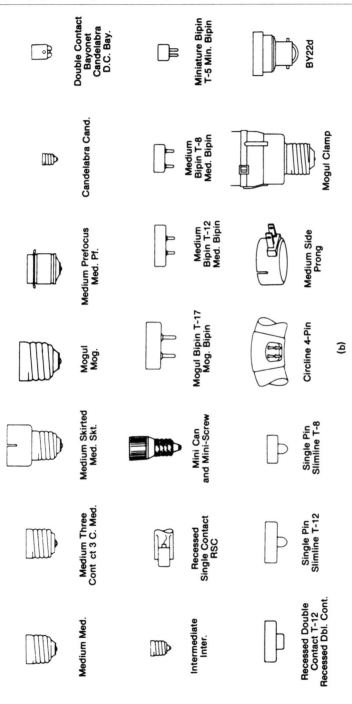

(b)

Figure 6-1b. (*Continued*). (a) Incandescent lamp; (b) common lamp bases; (c) common bulb designations. (*b, c, courtesy of Philips Lighting Corp.*)

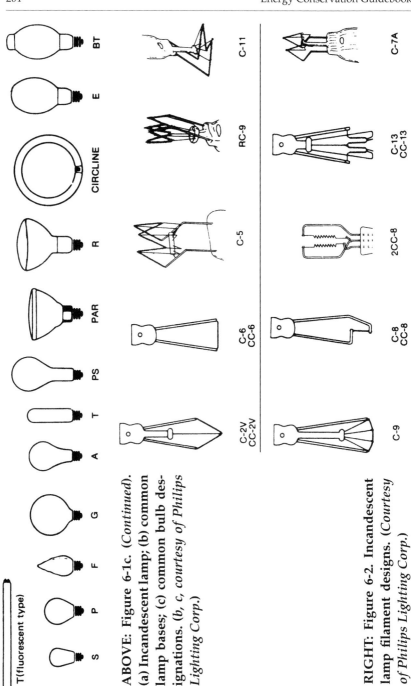

ABOVE: Figure 6-1c. (*Continued*). (a) Incandescent lamp; (b) common lamp bases; (c) common bulb designations. (*b, c, courtesy of Philips Lighting Corp.*)

RIGHT: Figure 6-2. Incandescent lamp filament designs. (*Courtesy of Philips Lighting Corp.*)

incandescent lamp is manufactured, the air is removed from the glass envelope to prevent the filament from burning, and an inert gas is added.

As lights get older, their output is reduced. Typically, when an incandescent lamp "burns out," its light output is approximately 85% of its original illumination. A decrease in the voltage of the power system also reduces the light output. A 1% decrease in voltage will cause an approximate 3% decrease in lighting output.

Because of the light from an incandescent lamp is a result of the glowing hot filament, this produces heat, an often unwanted by-product in many lighting applications. In some cases, the heat output of the lamp is so significant that the bulb is considered a heat source that also produces light. The temperatures can reach levels so high that they can cause severe burns to human skin and ignite combustible materials if brought in close contact. Also, the service life of such lighting sources is rather short because of the extremely high temperatures. However, incandescents do enjoy a few advantages over other lighting types. Incandescent lamps are relatively inexpensive, readily available, do not require any supporting circuitry, (such as a ballast) and can be used with the simplest of controls. Unfortunately this combination of light and heat as outputs makes this style of lamp rather inefficient compared to other types of lamps and lighting discussed in this chapter.

The basic design of incandescent lighting has remained relatively unchanged since the days of Thomas Edison, but improvements in filaments, gasses surrounding the filaments, lamp enclosures, and reflectors have helped improve the inefficient design. One such improvement is in the tungsten-halogen lamp design. One of the factors that limit service life of an incandescent lamp is filament life. Because it is the glowing hot filament that makes light in an incandescent lamp, techniques have been employed to help increase the life of the filament. Aside from the traditional evaporation technique, some lamps are filled with an inert gas, such as argon, krypton, or nitrogen. Similarly, tungsten-halogen lamps have a bulb filled with halogen gas, or a gas from the same chemical family. However, instead of burning up the filament as it glows and begins to break down, the gas in the lamp reacts with the spent tungsten, and allows the spent tungsten to recombine in a cooler area of the filament. While the efficiency of such lamps are 15% to 30% better than a conventional incandescent lamp, the cost can be as much as 50% higher.

FLUORESCENT LIGHTING

Fluorescent lighting such as that in the room shown in Figure 6-3 is used extensively in industrial and commercial buildings. Fluorescent lamps are tubular bulbs with a filament at each end. There is no internal electrical connection between the two filaments. The tube is filled with mercury vapor and when an electrical current flows through the two filaments a continuous arc is then formed between

Figure 6-3. Fluorescent lighting. (*Courtesy of Armstrong Cork Co.*)

them by the vapor. High-speed electrical particles passing between the filaments collide with the mercury atoms and produce ultraviolet radiation. The inside of the tube has a phosphor coating which reacts with the ultraviolet radiation to produce visible light.

The circuit for one type of fluorescent light is shown in Figure 6-4. Note that a thermal starter (bimetallic strip) and a heater are connected in series with the filaments. The bimetallic strip remains closed long enough for the filaments to heat and vaporize the mercury in the tube. The bimetallic switch will then bend and open, due to the heat produced by the electrical current through the heater. The filament circuit is now opened. A capacitor is connected across the bimetallic switch to reduce contact sparking. Once the contacts of the starter open, a high voltage is momentarily placed between the filaments of the lamp due to the action of the ballast coil. The ballast coil has many

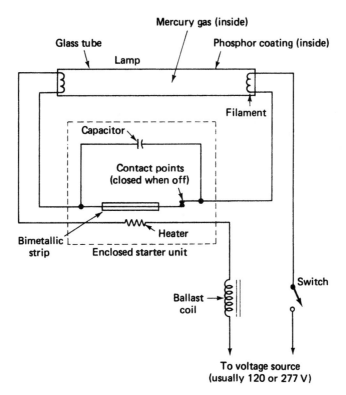

Figure 6-4. Fluorescent lighting circuit.

turns of wire that produce a high-voltage surge when the contacts of the starter separate. This effect is sometimes called inductive kick-back. The high voltage across the filaments causes the mercury to ionize and initiates a flow of current through the tube. Several methods are used to start fluorescent lights, but this method best illustrates the basic operating principle.

Fluorescent lights produce more light per watt than do incandescent lights and in the long run are cheaper to operate. Since the illumination is produced by a long tube, there is also less glare. The light produced by fluorescent bulbs is similar to natural daylight. The light is whiter and the operating temperature is much less with fluorescent lights than with incandescents. Various sizes are shapes of fluorescent bulbs are available. These sizes are usually expressed in eights of an inch in diameter. The two common bulb sizes are T-12 and T-8. A T-12 bulb is 1-1/2 or 12/8-in. diameter. Common lengths are 24-, 48-, 72-, and 96-in.

Fluorescent Light Ballasts

Ballasts are devices used to cause fluorescent lights to start and are important to the operation of the entire lighting system. A ballast is normally an enclosed coil or wire connected into the electrical circuit of a lamp. The ballast supplies the necessary voltage surge to develop an arc discharge to ionize the mercury gas within the lamp. A ballast also limits the current flow through the lamp and acts as a protective device to prevent the destruction of the lamp. There are several types of ballasts used with fluorescent systems. One particular model is shown in Figure 6-5.

Rapid-Start Circuit

The rapid-start or trigger-start ballast has become the most popular type of ballast for domestic use in the United States. This particular ballast contains transformer windings that constantly give the required voltage and current for electrode heating. A schematic diagram of this lamp circuit is shown in Figure 6-6. When the circuit is energized, the windings quickly heat the electrodes which then allows the tube for the lamp to arc. The ballast is smaller and there is no flickering associated with this type of circuit.

The continuously heated cathode of the rapid-start lamp is bet-

Figure 6-5. Fluorescent lamp ballast. (*Courtesy of Lutron Electronics Co., Inc.*)

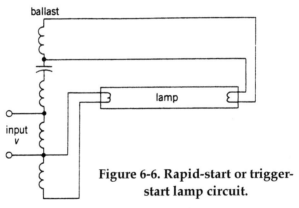

Figure 6-6. Rapid-start or trigger-start lamp circuit.

ter adapted to higher lamp currents and wattages. Rapid-start lamps are available in sizes up to 215 watts with smaller diameters. Special circuits are available that permit the owner to dim this particular fluorescent lamp. Rapid-start lamps are commonly used in flashing signs, where continuously variable illumination levels are needed.

High-Efficiency Ballasts

Due to their design, typical installations of fluorescent lamps were limited to simple on-off controls. While effective and inexpensive, these types of controls are very limited. As technology evolves, controls for fluorescent lamps have evolved. Dimming ballasts based in digital electronics are now commercially available that allows the intensity of fluorescent lamps to be altered. This digital technology is the basis for a new breed of ballast, the high-efficiency ballast that offers programmability in starting options as well as intensity. Such high-efficiency ballasts not only save up to 30% in energy over traditional fluorescent systems, can control six or more lamps (magnetic ballasts are limited to two lamps), and thereby decreasing costs in fixtures. Other advantages of the newer digital ballasts include reduced hum and reduced lamp flicker. When these advantages are combined with the energy savings and ability to dim, users are better able to meet their lighting needs with this generation of ballast. As federal legislation moves to enact stricter efficiency standards for fluorescent lighting ballasts, magnetic ballasts will no longer be available. Initially the drawback to changing from the magnetic-type ballasts will be cost, however as with any technology, as more products such as these are brought to market, the price will decline. In the future, these digitally designed ballasts can be made compatible with computer networks that will control the lighting for entire facilities, possibly from a remote location.

VAPOR LIGHTING

As you look upward in a gymnasium, parking lot, sports arena, or an expansive retail store, you are likely to encounter an example of lighting from the family of vapor lights. By definition, these lamps are able to produce light by arcing in the presence of a pressurized gas of mercury or sodium. These lamps produce a large volume of light, suitable for outdoor or very large indoor applications. Typically lamps in from the vapor lighting classification do not reach full intensity when switched on, but may take several minutes to 'warm up'. Other unique characteristics from the vapor group of lights are the color of the emitted light. Golds, yellows, blues, and greens are but some of the colors from a vapor lamp, while other types of lighting produce the traditional white or daylight color. While they are more energy efficient than traditional incandescent lamps to produce the same amount of light, their initial cost is much higher, as well as maintenance costs such as bulb replacement.

High Intensity Discharge (HID) Lighting

Another popular form of lighting is the vapor type. The mercury-vapor and the sodium-vapor light are two of the more common types in this division. These lights are filled with a gas that produces a characteristic color. For instance, mercury-vapor produces a greenish-white light and argon gives off a bluish-white light. Gases may be mixed to produce various color combinations as is often done when used for advertising purposes.

A mercury-vapor lamp is shown in Figure 6-7. It consists of two tubes with an arc tube placed inside of an outer bulb. The inner tube contains mercury. When a voltage is applied between the starting probe and an electrode, an arc is started between them. The arc current is limited by a series resistor; however, the current is enough to cause the mercury in the inner tube to ionize. Once the mercury has ionized, an intense greenish-blue light is produced. Mercury-vapor lights are compact, long-lasting, and easy to maintain. They are used to provide a high-intensity light output. At low voltages, mercury is slow to vaporize, so these lamps require a long starting time (some-

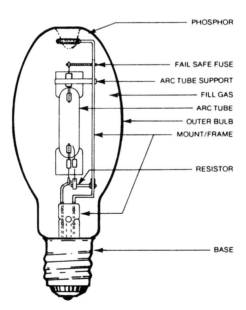

Figure 6-7. Mercury-vapor lamp. (*Courtesy of North American Philips Lighting Corp.*)

times 4-8 minutes). Mercury-vapor lights are also used for outdoor lighting. An application of the mercury-vapor lighting is shown in Figure 6-8.

The sodium-vapor lamp of Figure 6-9 is popular for outdoor lighting and for highway lighting. When an electric current passes through sodium, electrons are given off. The ionizing circuit uses a positive charge which is placed on the electrodes. As electrons pass from the heater to the positive electrodes, the neon gas is ionized. The ionization of the neon gas produces enough heat to cause the sodium to ionize. A yellowish light is produced by the sodium vapor. The sodium-vapor light can produce about three times the candlepower per watt as that of an incandescent lamp.

There are several types of high-intensity-discharge (HID) lamps in use today. These are a clarification of lamps which produce light when a high-voltage arc passes through a vapor-filled tube. Until re-

Figure 6-8. Application of mercury-vapor lighting. (*Courtesy of Philips Lighting Corp.***)**

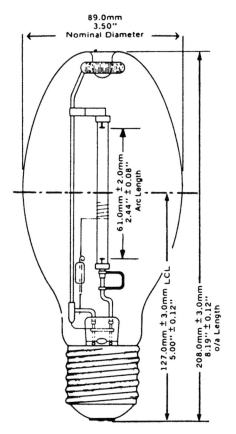

Figure 6-9. Sodium-vapor lamp.
(Courtesy of Philips Lighting Corp.)

cently, HID lamps were used in industries where color qualities were not important. They now have improved color characteristics and are available in a variety of sizes for both indoor and outdoor use.

Metal-halide Lamps

Metal-halide lamps are essentially mercury-vapor lamps that have been altered by the addition of different compounds in the enclosed portion of the lamp. The color characteristics of metal-halide lamps are different from those of mercury-vapor lamps.

Ballasts and Igniters

Similar to florescent lamps, HID lamps require a ballast or igniter for the starting and control of current in the lamp. Ballasts for HID lighting vary from a simple coil or transformer to complex units involving capacitors and autotransformers. Just as ballasts for florescent lighting are being shaped by technology, so are HID ballasts. Digital technology allows for the next generation of HID ballasts to be more efficient, control more lamps, and to allow for centralized computer controls, much like their counterparts for fluorescent lighting.

STREET LIGHTING

Some modern street lights are shown in Figure 6-10. The lighting systems of today are highly reliable compared to the systems that were installed many years ago. Earlier systems were either turned on or off manually or by timing devices that were regulated by the time of day rather than by the natural light intensity. Some were controlled

Figure 6-10. Street lighting. (*Courtesy of Philips Lighting Corp.***)**

by electrical impulses that were transmitted on the power lines. Now most systems are controlled by an automatic photoelectric circuit. The lights now used to illuminate streets and highways have photoelectric controls. They operate during periods of darkness and are automatically turned off when natural light is present.

The earliest types of street lights used were 200- to 1000-watt incandescent lamps. Now mercury and sodium-vapor lamps are the primary types used for roadways. Mercury lamps produce a white light and sodium lamps give off a yellowish color. Several different lamp designs and mounting fixtures are also available with these two particular types.

Several years ago, street lights were converted from incandescent to mercury-vapor lamps. The trend now seems to be toward the use of sodium-vapor lights. Several areas of the country have converted to the use of sodium-vapor lamps. They tend to produce more illumination than a similar mercury-vapor light. Less electrical power to produce the same amount of illumination is also required. The ability of sodium-vapor to deliver more light with less power consumption makes them more economically attractive.

LED LIGHTING

In an effort to reduce energy consumption, replacing traditional incandescent lamps with units using light emitting diodes, or LEDs is an emerging technology that shows a lot of promise. In addition to lower power consumption, the LED lights have a longer life than incandescent bulbs, emit much less heat, and have the ability to point or spotlight where light is need. At the time of this book, LED lighting is not a cost effective replacement, but as the technology evolves, and production levels increase, costs may drop to a level where it is a viable alternative in the near future. Some examples of LED lamps are shown in Figure 6-11.

Most people have encountered applications where LEDs have already been used to replace incandescent lamps. Tail and signal lamps in personal and commercial vehicles are beginning to use this technology, it is found in novelty or seasonal decorations, many outdoor landscaping lights use LEDs for a light source, and traffic signals

Figure 6-11. Screw-base LED lamps. *(Photos courtesy of LEDtronics, Inc.)*

benefit from the long-life of the LED lamps. Also, if you have used fiber optic cables for data or voice transmission, the device that generates the light, a modulator, was likely a LED. Because of their long life, ruggedness, and low power consumption LEDs offer, many flashlight manufactures are now selling units that use LEDs or multiple LEDs as a light source, rather than an incandescent bulb.

LEDs as Light Sources

The basis for LED lights is the LEDs themselves. As the name implies, the light is emitted from semiconductor electronic devices called diodes. While typical diodes do not emit light, but are uses to control current flow, special classes of diodes are manufactured specifically for the production of light, both within and outside the visible light range. LEDs are made of aluminum-gallium-arsenide (AlGaAs), a material that has the ability to produce light at the atomic level. In the LED's atomic structure, particles called photons, the most basic units of light, are present. These photons are released when current passes through the diode, causing electrons to change orbits around the nucleus. The width of this gap between the orbits is a factor in determining the wavelength (color) of light emitted, and the energy necessary to produce light. Because the diodes are made from aluminum-gallium-arsenide, rather than germanium or silicon, they produce light from the emission of the photons. Below is a list of some of the advantages and disadvantages of LED lighting over conventional methods.

Advantages
- Draws as little as 1/10 energy as much as incandescent lamp
- Draws as little as 1/4 energy as much as a fluorescent lamp
- Lasts 10 times longer (or more) than conventional lighting
- Prices are steadily decreasing
- No stray radio frequency interference for noise sensitive environments
- No flicker, a constant light source
- Packages are smaller than compact fluorescent
- Produces little heat
- Can be made into packages not available to incandescent and fluorescent, such as strips
- Ability to withstand shock and vibration

Disadvantages
- Requires use of transformer and other electronic components
- Much higher cost
- Not available in all standard configurations

Many styles and types of LED lighting are available. Designs to light areas such as desktops or parking lots in the form of flood lamps or spot lamps are available. Also, signage, walkway, and landscape configurations are readily available. While obtainable, they do have a drawback. The LED lights currently available to the public require the replacement of the fixture as well as the light source. An LED lamp to directly replace an incandescent lamp screw base is available, but it is not currently cost effective. As the technology is improved, a suitable unit at a competitive price will reach the market.

The primary drawback of using LEDs to replace standard lighting is cost. While the initial cost of an LED lamp is much higher, if all the energy factors related to lighting over a time period puts the initial higher cost in a new perspective. For example, a replacement incandescent lamp may only cost a few dollars, where the LED replacement may be over fifty dollars, or nearly one hundred dollars. However, if you take into account the energy savings over years of use, and include the labor factor of replacing ten or more incandescent lamps for every one LED lamp, the payback in commercial applications can be achieved in a few years. Vendors that sell the LED lamp products have energy calculators available that demonstrate the savings over time.

LIGHTING DESIGN

The lighting design of a building should include consideration of initial cost, maintenance cost, energy use, and appropriateness for the use. A lighting system in any part of a building should create the most proper and pleasing lighting at the most economic price. Several factors affect the type of a lighting system that is used for a particular application. Among the questions which should be considered are:

1. Is the lighting adequate?

2. Is the lighting equipment (luminaires, covers, etc.) appropriate for the application?

3. Does the lighting provide the proper quality of light?

4. How much energy does the lighting system use?

The existing lighting system design of a building should be analyzed to see if there are ways to reduce energy use without adversely affecting the lighting design. There are several methods of reducing the energy use of lighting systems listed in the checklist at the end of the chapter.

Lighting Circuits

For electrical lighting control there are several different types of circuits. Several of the circuits used for incandescent, fluorescent, and vapor lighting systems will be studied. Electrical lighting circuits for different areas of buildings must be wired properly. Some lighting fixtures are controlled from one point by one switch, while other fixtures may be controlled from two or more points by a switch at each location.

These switches or lighting controls should be mounted so they are easily accessible by the user and clearly visible. In commercial applications, lighting controls should be well-labeled so individuals unfamiliar with the structure can control the lighting properly.

Lighting circuits are simple electrical circuits. One common type of lighting circuit is a 120-volt branch that extends from a power distribution panel to a light fixture or fixtures in some area of a building. The path for the electrical distribution is controlled by one or more switches that are usually placed in small metal or plastic enclosures inside a wall. These switches are then covered by rectangular plastic plates to prevent possible shock hazards.

Switches as previously described are part of a classification of controls known as general controls. These switches can be individual, part of a switch gang, or can be controlled by a key, a pull cord or a dimmer. These general controls are titled as manual controls because someone or something must operate them manually. There is another classification of general controls known as automatic. In this classification, timers, motion sensors, or photo controls are added to lighting circuits to allow them control the lighting independent of any manual input. The most common example of an automatic control is the photo sensor enabling street lights to operate from dusk-to-dawn with no manual intervention.

The design of lighting circuits involves calculation of the maximum electrical current that can flow through the lights connected to a circuit. The National Electrical Code specifies a minimum requirement of lighting for various types of buildings. A minimum number of watts of light per square foot is required for each type of building.

Lighting-Fixture Design

Lighting fixtures are the units used to hold lamps in place. They are technically referred to as luminaires. Luminaires are used to efficiently transfer light from its source to a work surface. The proper design of luminaires allows an efficient transfer of light. It is important to keep in mind that light intensity varies inversely as the square of the distance from the light source. Therefore, if the distance is doubled, the light intensity would be reduced four times.

Many factors must be considered in determining the amount of light that is transferred from a light bulb to a work surface. Some light is absorbed by the walls and by the light fixture itself. In other words, not all light is efficiently transferred. The manufacturers of lighting systems develop charts that are used to predict the amount of light that will be transferred to a surface as shown in Figure 6-12. These charts consider the necessary variables for making a prediction of the quantity of light falling onto a surface.

Each luminaire has a rating that is referred to as its coefficient of utilization (CU). The coefficient of utilization expresses the percentage of light output that will be transferred to a work area. The coefficient of a luminaire is determined by laboratory tests made by the manufacturer. These coefficiency charts also take into consideration the light-absorption characteristics of walls, ceilings, and floors. The charts do this when determining the coefficient of utilization of a luminaire.

Another factor used for determining the coefficient of utilization is called the room ratio. Room ratio is simply determined by the formula:

$$\text{room ratio} = W \times L / H(W + L)$$

where,

W is the room width, in feet

L is the room length, in feet

H is the distance from the light source to the work surface, in feet

Note: Work surfaces are considered to be 2.5 feet from the floor, unless otherwise specified.

Factors in Determining Light Output

There are several other factors that must also be considered in determining the amount of light transferred from a light source to a work surface. Through experience, we know that the age of a lamp has an effect on its light output. Lamp manufacturers determine a depreciation factor or maintenance factor for luminaires. This factor expresses the percentage of light output available from a light source. A depreciation factor of 0.75 means that in the daily use of a light source, only 75% of the actual light output is available for transfer to the work surface. The depreciation factor is an average value. It considers the reduction of light output with age and the accumulation of dust and

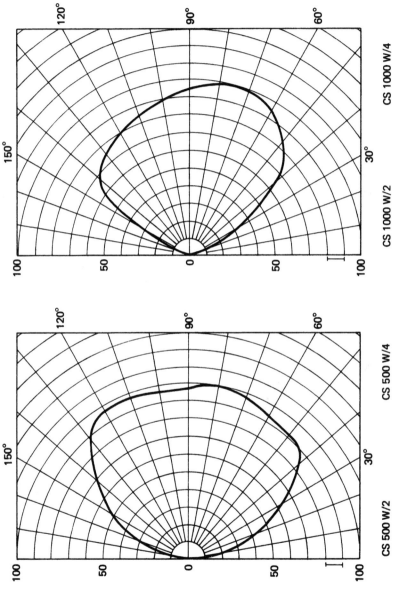

Figure 6-12. Luminous intensity (light distribution) charts. (*Courtesy of Philips Lighting Corp.*)

dirt on the luminaires. Some collect dust more easily than others.

The following problem shows the effect of the CU and the depreciation factor (DF) on the light output transferred to a work surface.

1. Given — A lighting system for a building has 20 luminaires. Each
    luminaire has two fluorescent lamps. Each lamp has a light out-
    put of 2000 lumens. The CU and the DF found in e chart devel-
    oped by the manufacturer are 0.45 and 0.90, respectively.

2. Find — The total light output from the lighting system.

3. Solution:
    a. Find the total light output from the lamps:
        2000 lumens per lamp × 10 lumens × 2
        lamps per luminaire = 40,000 lumens.
    b. Find the total output:
        light output = total lumens × CU × DF
        = 40,000 × 0.45 × 0.90
        = 16,200 lumens

The distribution of light output onto work surfaces must also be
considered. The CU and the DF are used to find the total light output
of a lighting system. A greater light output is required for larger areas.
The light output to a work surface is expressed as lumens per square
foot, or footcandles. A light meter may be used to measure the quantity of light that reaches a surface.

The following two formulas are useful for finding the required
lumens for a work area or the light output from a particular lighting
system:

total lumens = desired footcandles × room area (ft)/CU × DF,
and

footcandles available = total lumens × CU × DF/room area (ft).

Types of Illumination

The basic types of illumination in a building include general,
combined, and local lighting. General lighting provides illumination
throughout a particular area. Lamps are ordinarily uniformly spaced
in the ceiling to provide general lighting. Local lighting is used to

provide higher levels of illumination for specific tasks. Combined systems are used where the general visual tasks require a lower level of light but there is need for local lighting of high intensity. The type of illumination actually needed in each part of a building should be considered in determining the actual lighting needs for each individual area. There are many ways to reduce lighting in buildings to conserve energy and save money on electric bills.

Illuminating Engineering Society

An organization for those interested in lighting systems is the Illuminating Engineering Society (IES). This group publishes a lighting handbook which can be used for the design of lighting systems and develops standards for lighting utilization.

LIGHT DIMMING

Most light sources can be dimmed by reducing the electrical power supplied to them. Light dimmer controls allow the level of light in a room to be varied over a wide range of illumination. This control can be used to adjust light level according to need, such as time of day, type of room, and activity being performed. When used properly, light dimmers can conserve electricity and increase the life of lamps. Care should be taken to turn dimmers off when they are not in use since dimmer circuits also use energy even when the lights are turned off.

Manual and automatic are the two types of dimmer controls. Manual controls are adjusted by the user to the desired level of light output. Automatic systems adjust the level of light by using photo-electric sensors to control power applied to the lamp according to need. Dimmer switches may be purchased to replace standard lamp switches.

Effect of Reduced Lighting on Heat

The electrical energy supplied to lamps not only produces light, but also produces a great deal of heat. During the cooling season, the heat produced by lights causes an additional load on the air-conditioning system of the building. When reduced-wattage lighting is used, the heat is reduced in the area and the air-conditioning load is reduced. This energy reduction causes a savings in lighting as well as for air conditioning.

On the other hand, the effect of heat produced by lamps during the heating season reduces the load of the heating system. Reduced lamp wattage means that less heat is available in a building. The heating system must supply more heat to compensate for this loss.

The precise energy savings brought by reducing the wattage of lamps is a complex calculation. Several factors, such as climate, hours of lamp operation, type of HVAC system, and condition of the building, must be considered in determining energy savings through reduced wattages of lighting systems. In most situations, the probability of energy savings and reduced energy costs brought about by lower-wattage lamps is high.

TIPS FOR ENERGY CONSERVATION
IN LIGHTING SYSTEMS

1. Use the most efficient light source available, since more efficient lamps produce a greater amount of light per watt.

2. Use heat-transfer luminaires to utilize heat from lights in cold weather and to remove heat in warm weather.

3. Always turn lights off when they are not needed.

4. Use luminaires effectively to provide maximum light for the task performed.

5. Plan proper lighting system maintenance for keeping lights clean and lighting levels adequate.

6. Design lighting systems for anticipated activities in each area of the building.

7. Use daylight hours to maximum advantage so that electric lighting may be reduced.

8. Use lighter colors on ceilings, walls, floors, furniture, and equipment to increase reflectance of light.

Modification of Lighting System

The modification of existing lighting systems should be considered to provide energy conservation and financial savings. Most buildings have areas that are overly illuminated. The simplest and most cost-effective way to reduce the light level is to remove lamps. This can be done without making an area appear less attractive. The *IES Handbook* should be checked to determine minimum lighting levels. It is possible to estimate energy and money savings by removing lamps. One manufacturer has developed the nomograph shown in Figure 6-13 to estimate energy savings due to lamp removal.

Energy can also be saved by modifying existing light sources by replacement with another source. For example, incandescent lamps could be replaced with more efficient fluorescent lamps. Initial costs, maintenance costs, and life expectancy of lamps should be considered before replacing to accomplish energy conservation. The estimated lumens per watts of output of several types of light sources is shown in Figure 6-14.

PRODUCTS FOR ENERGY CONSERVATION

Replacements for Incandescent Lamps

Several manufacturers have produced lamps that can be used to replace standard-base incandescent lamps. The unit shown in Figure 6-15 is referred to as a Powermiser. This unit is a mercury-vapor lamp which fits into a standard 120-volt incandescent lamp fixture. The conversion from incandescent to mercury-vapor lighting requires no rewiring or modifications. The ballast for this system is mounted internally. These units should provide reduced energy consumption.

Another replacement for incandescent lamps is shown in Figure 6-16. This unit is a circular fluorescent lamp which screws into incandescent lamp bases of portable or ceiling lamps. A fluorescent lamp produces more light than an incandescent lamp of the same wattage rating. In other words, energy costs can be reduced without sacrificing light output levels.

ENERGY EQUALS MONEY—

Here is a conversion chart that tells you how much you will actually save in dollars based on the geographical costs of kWh and as a function of square footage of your building.

INSTRUCTIONS FOR NOMOGRAPH

1. Locate number of lamps to be removed on scale A.
2. Place straightedge from number of lamps to hours of operation per year of scale E.
3. Mark intersection with scale C. This is the number of kWh that will be saved per year, including air conditioning.
3. Place straightedge from mark made in step 3 to effective cost per kWh on scale F.
4. Place straightedge from mark made in step 3 to effective cost per kWh on scale F.
5. Mark intersection with scale D. This is the approximate dollar savings per year realized by removing the lamps.

COMMENT:

1. If the number of lamps to be removed is more than 1000, divide the number by 10 and use the nomograph. Then multiply the savings by 10. If the number of lamps is less than 100, multiply the number by 10, use the nomograph, and divide the savings by 10.
2. Kilowatts saved (scale B) includes an allowance of 1/3 watt of air conditioning for each watt of lighting, which is typical of medium-sized installations.
3. Effective cost per kWh (scale F) should be determined by allowing for fixed charges, and any other charges made by the utility, the whole divided by kWh used to get the effective cost per kWh.

EXAMPLE:

600 lamps are to be removed from a lighting system. The system is operated 2000 hours/year, and the effective cost per kWh is 7¢. Connect 600 on scale A with 2000 on scale E. The intersection with scale C is at 72,000 kWh, which is the amount saved per year. From this intersection, run a line (B) to 7¢ on scale F. The intersection with scale D is at $5000, which is the dollar amount that will be saved in energy per year.

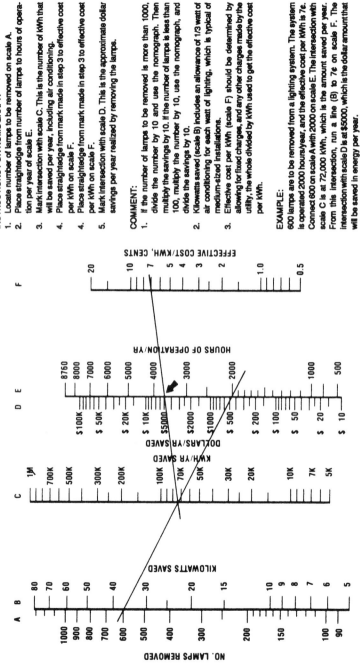

Figure 6-13. Nomograph of savings through lamp removal. *(Courtesy of Polarized Corporation of America.)*

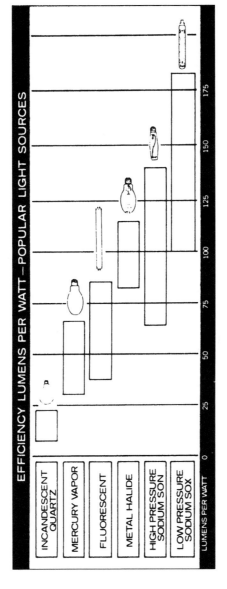

Figure 6-14. Lumens per watt output of popular light sources. (*Courtesy of North American Philips Lighting Corp.*)

Figure 6-15. Crouse-Hinds Powermiser used to replace incandescent lamps. (*Courtesy of Crouse-Hinds Co.*)

Compact Fluorescent Lamp

A long-life alternative to the incandescent lamp is the compact fluorescent lamp. Originally titled the E-lamp, this design combining the lower operating costs of fluorescent lamps with the ability to retrofit existing incandescent fixtures has become very popular since the beginning of the 21st century. This miniature version of a fluorescent lamp contains a specially shaped fluorescent bulb in a biax or spiral configuration, miniature ballast, and a standard screw base all in an area similar to the incandescent bulb it is designed to replace. Because of all these added components, the cost of a compact fluorescent is estimated to be at least four times as expensive as a traditional incandescent lamp.

The life of a compact fluorescent lamp is similar to those expected by typical fluorescent lights, but to 20,000 hours, twenty times the

Figure 6-16. Circular fluorescent unit used to replace incandescent lamps. (*Courtesy of General Electric Company.*)

expected life of an incandescent lamp. Because compact fluorescent lamps use only approximately 25% the amount of energy of their incandescent counterparts, they are much less expensive to operate. This savings is so significant it is estimated that a compact fluorescent lamp can save enough money in electricity costs to offset the initial higher price within about 500 hours of use, which is typically less than one year.

The lamp operates by, first, generating high-frequency radio signals. The rapidly oscillating radio waves excite a gas mixture in a sealed tube placed inside of the outer bulb. The moving gas, in the sealed inner tube, gives off light. When this light hits the main bulb's phosphor coating on the inside of the glass shell, visible light is then apparent. Since there is no electrode or filament, the lamp will only gradually dim until the phosphor wears out.

In addition to increased cost, the lamp has a few other draw-backs. Like traditional fluorescent lamps, they must 'warm up' in order to achieve optimal lighting. Also because of the unconventional design of the fluorescent tubes, the light output of a compact fluorescent can be considered somewhat diminished. If the lighting fixture was controlled by manual dimmer, the compact fluorescent cannot be dimmed in this manner. Also, the high-frequency radio wave generated by the bulb may generate radio frequency interference. This may cause other objects in the area that operate on similar frequencies to become inoperable or behave erratically. Lastly, because of the need for many internal components, the lamps are actually slightly larger than the incandescent lamp they are meant to replace, and my not fit in the fixture correctly, or be less appealing to the eye.

The compact fluorescent lamp does have advantages. The afore-mentioned long life and relatively short pay-back time is certainly among the most attractive features of the lamp. If adopted on a large scale, the savings to the nation could be relatively significant. Also, because fluorescent lamps do not produce the heat that is generally associated with incandescent lamps, there will be reduced demands on summer heating loads. Also, because of the long-life of the lamp, it will be easier and more convenient to maintain because of the need for fewer replacements. Despite these apparent advantages, it appears the public has been slow to transition from incandescent to compact fluorescent lamps.

Fluorescent Lighting Controllers

A fluorescent lighting controller is a solid-waste power control unit. This piece is used to reduce energy consumption of fluorescent lights by reducing lighting levels. An example of this unit is shown in Figure 6-17. Installation would be simple and the unit would not require any special ballasts or lamps. Many buildings have exceptionally high lighting levels and could save energy by reducing light levels. The manufacturer of this unit also markets a light sensor such as the one shown in Figure 6-18. This is a photoelectric unit which mounts into the ceiling and senses light output. The amount of light produced at a certain area can be continuously monitored. A constant light output can be obtained by using a light sensor to adjust the power supplied to a fluorescent fixture.

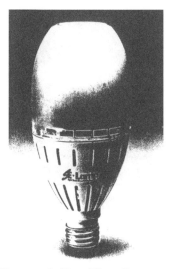

Figure 6-17a. The Compact fluorescent lamp. (*Courtesy Intersource Technologies Inc.*)

Figure 17b. Typical compact fluorescent lamps.

Special-effect Lighting Fixtures

It is possible to modify light output by using special-effect lighting fixtures. One such fixture is shown in Figure 6-19. This fixture is used to distribute fluorescent light though a prismatic lens system to cause light to disperse over a large area and produce glare-free light. The manufacturer claims that significant energy savings can result through the use of these fixtures.

Another special-effect fixture is shown in Figure 6-20. These are fluorescent luminaires that use anodized aluminum baffles as a means of diffusing light. They are used to distribute light more evenly and more economically.

Another type of unit that can be considered for interior use in industrial and commercial buildings is shown in Figure 6-21. This luminaire uses a special design to direct light output to a work area. The unit has a prismatic light refractor which distributes light evenly at wide angles on both vertical and horizontal surfaces, producing uniform illumination. The luminaire shown in Figure 6-4 also produces uniform lighting, but its design is recessed.

Figure 6-18. Compact fluorescent lamps in an incandescent fixture.

Automatic Lighting Control

Lighting is a major energy expense for many types of buildings. A way to reduce lighting costs is to use an automatic lighting control such as the one shown in Figure 6-22. These controllers can be used to turn off lights when they are not needed and to lower lighting levels in certain areas of the building when the area is not being utilized. Many lighting controllers are microprocessor-based systems which provide fully automatic operation of lighting systems and are easy to program. The obvious advantage of using automatic lighting control is that it places the control in the hands of the employees. Many types of buildings could use automatic lighting control to save large amounts of money on energy costs.

High-mast Lighting

High-mast lighting is a system that uses several luminaires mounted on higher poles to light a large area. They may be used for highway lighting and other outdoor uses, such as large parking lots at other industries and shopping centers and freight terminals. The advantages of high-mast lighting include the use of fewer poles, lower installation cost, reduced operating costs, and more energy efficiency. The light distribution characteristics of high-mast systems make their use desirable for many outdoor applications.

Figure 6-19. Fluorescent retrofit power control module. (*Courtesy of Lutron Electronics Co., Inc.*)

Figure 6-20. Photoelectric light sensor. (*Courtesy of Lutron Electronics Co., Inc.*)

Figure 6-21. Special-effect lighting fixtures. (*Courtesy of American Optic-Lite Corp.*)

Self-Powered Signs

Large buildings use many "EXIT" signs near doors and hallways. These are ordinary powered by the building electrical system. Self-powered signs such as the one shown in Figure 6-23 provide reliable exit identification without consuming any energy.

ENERGY CONSERVATION CHECKLIST
FOR LIGHTING SYSTEMS

Table 6-1 on page 227 presents a checklist for lighting systems.

Figure 6-22. Special-effect fluorescent luminaires. (*Courtesy of USI Columbia*)

Figure 6-23. "Superwatt" luminaire. (*Courtesy of The Miller Co.*)

Figure 6-24. Automatic lighting controller. (*Courtesy of Square D Company*)

Figure 6-25. Self-powered exit sign. Self-powered signs provide reliable exit identification, day or night, without consuming any energy. (*Courtesy of Self-Powered Lighting Limited.*)

Table 6-1: Checklist for Lighting Systems/Improved Efficiency

Items to check	Corrective action
Lighting utilization. Check the building to see if lights are turned on in unoccupied areas. See if enough switches are provided for individual control of lights.	Use signs in rooms to instruct occupants to turn lights off when the room is vacant; organize work areas so that unnecessary lighting is not used; rewire light switches so that one switch does not control a large lighted area; use timers or photo-electric control to assure that lights are not used unnecessarily.
Excessive lighting levels. Check with a light meter to see if proper amounts of light are provided for each area of the building.	Turn off or disconnect lights as needed to accomplish proper illuminating; turn off lights near open areas in daylight periods; replace present lights with lower-wattage lamps; use portable lights such as desk lamps for individual use of a room.
Outside lighting. Check outside lights to make sure that they are used only for safety and security purposes when needed.	Stop using outdoor lights that are not needed; replace present lights with lower-wattage lights that provide adequate illumination; install timers or photoelectric controls to turn off lights when not needed; replace incandescent lights with more efficient types, such as sodium-vapor or metal-halide.

Table 6-1: Checklist for Lighting Systems/Improved Efficiency (*Continued*)

Items to check	Corrective action
Use of incandescent lights. Check the types of lighting used in each part of the building and the wattage of each light.	Use one larger incandescent lamp rather than two or more smaller lamps when possible; replace incandescent lights with more efficient fluorescent or mercury-vapor lamps.
Use of fluorescent lights. Check fluorescent fixtures to see if lighting level can be reduced by removing lamps.	When possible, remove two lights and ballasts from four lamp fixtures; in low-use areas, remove all fixtures possible to maintain adequate lighting level .
Fluorescent ballasts. Check fluorescent fixtures to determine if lamps that have been removed still have ballasts connected.	Have an electrician disconnect all ballasts that are not in use.
Cleanliness of lighting fixtures. Check lights and light fixtures to see if dirt has accumulated or shields have become discolored.	Inspect and clean lamps and fixtures on a scheduled basis to avoid inefficient lighting; replace shields that have become discolored; replace old or dam-aged luminaires with modern types that are easy to clean .
Replacement of lamps. Check the wattage of lamps and ballasts that have burned out to assure that minimum wattage is used.	Replace fluorescent lamps with more efficient, lower-wattage units; replace ballasts with lower-wattage units.

Chapter 7

Water Systems/
Saving Our Valuable Resource

INTRODUCTION

Nearly all modern buildings are equipped in some way with a water system. The life of all living things is based to some extent upon an adequate supply of drinking water. As a rule, people need only a small amount of drinking water to actually sustain life. A supply of drinking water is only one of the primary functions of a water system. In addition to this, water is needed for bathing, cleaning, cooking, washing clothes, and waste disposal. American water systems produce enough water each day to supply every person with approximately 150 gallons for general use.

In primitive circumstances, people would take water from ponds, streams, or rivers and either drink it on the spot or carry it in containers to another location. Most of the water that is available to us today is through surface streams or ponds, and is unsafe for human consumption. It must be cleared of chemical and bacterial contamination of significant value. It is then piped under pressure from the primary source to respective building sites, where it is used.

WATER-SYSTEM BASICS

Water systems follow the same basic format that we have described for other building systems. This means that in order to function properly, the system must have an energy source, transmission path, control, a load device, and possibly one or more indicators. These parts are an absolute necessity for satisfactory operation.

The individual parts of a water system are quite unusual compared to those of other systems. In practice, these parts may be situated at rather remote locations. A building, for example, represents the end result or load of a water system. All buildings supplied by the same primary source of water form a composite load for the entire system. Work is accomplished by the load when water is delivered to a selected place at a desired time. The primary source of water, which is usually a lake or river, may be located several miles from the area where it is actually being used. A network of pipes serves as the transmission path. System control is accomplished by cutoff valves, flow regulators, and anything that alters the flow of water. Indicators appear at numerous places throughout the system. Water meters are placed in the service connection of each building to monitor water consumption. The components of a simplified water system are shown in Figure 7-1.

WATER PURIFICATION

A unique part of a water system that distinguishes it from all other systems is the purification function. In a strict sense, the term *pure water* applies only to the purpose that it serves. Industrial facilities that consume large quantities of water for cooling and other manufacturing operations may not be necessarily concerned with a high

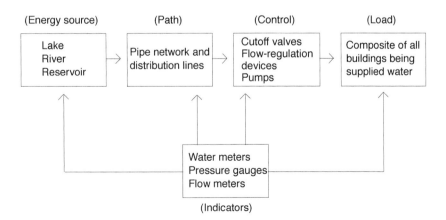

Figure 7-1. Components of a water system.

level of water purity. *Drinking* or *potable water* on the other hand, must be clear, cool, and free of any special taste or odor. It should contain very little or only a small amount of mineral salts. It should also be free of all harmful bacteria and chemical pollutants. Unless a special water feed system is installed, all water systems distribute potable water.

The potable water of a corporate water system normally goes through three unique purification operations before it is distributed for consumption. In simplified terms, this is achieved by coagulation and sedimentation, filtration, and disinfection. These processes are essential for water purification but can be achieved in a variety of ways, depending on the system, its primary source of water, and when it was constructed. Figure 7-2 shows a simplification of the water purification function.

Most city water systems achieve purification by the filtration method. Water from the primary source is pumped or forced into a chemical building or area for the first phase of the purification process. Water arriving in this area is mixed with prescribed quantities of lime, activated carbon, alum, and chlorine. From this area the mixture then passes into a coagulation basin, where it is thoroughly mixed. A coagulant is a specific type of chemical that forms a small fluffy mass when placed in water. Aluminum sulfate or fiber alum is a common chemical used as a coagulant. The fluffy mass formed by this action is called *floc*. Floc particles are used to remove foreign material suspended in the water by adhering to it. Floc and other suspended material settle to the bottom of a basin after a short period of time. This action is representative of the coagulation and sedimentation function of the purification process.

The filtration function is applied to water after it leaves the coagulation basin. Filtering is accomplished by passing water through a bed of sand approximately 30 in. (76 cm) thick. Water filters slowly down through the sand into an underdrainage area. Accumulated water in this area is then carried to a clear water-storage basin.

The disinfection function is designed to keep water from carrying infectious diseases. Most purification operations achieve this function by adding chlorine to water after it reaches the clear water-storage basin. A very small amount of chlorine is needed to disinfect quantities of water. One pound of chlorine can disinfect

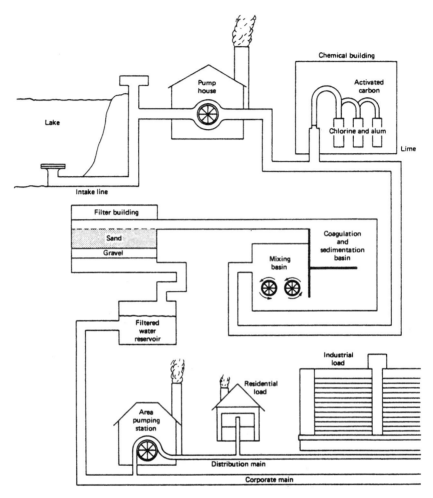

Figure 7-2. Composite city water system.

200,000 gallons of water. In addition to this, some systems may add a softening agent and fluoride to the water at this stage. The softening agent is designed to make water more suitable for home and industrial uses, and fluoride helps to reduce tooth decay. At this point water is pumped into the transmission network for distribution throughout the system.

WATER DISTRIBUTION

After the purification function has been achieved, water must be distributed to homes, commercial buildings, and industrial sites. The distribution function of the water system generally represents the most costly part of the entire system. Vast networks of pipes are used to distribute water from the source to the respective building sites of the system. The physical size and geographical location of the system has a great deal to do with the distribution function. In general, distribution includes pumping stations, street mains, reservoirs or standpipes (water towers), and building plumbing equipment.

The pumping station of the distribution network is designed to lift water to various heights and to develop enough force to pass it through the system. Electric-motor-driven pumps are commonly used to achieve this function. Fundamentally, the pump accepts water at its inlet port and forces it to move through a confined area before being expelled at the outlet port. Increasing the flow rate of water by pumping action causes it to develop pressure as it moves through the system. A variety of different pumps are available to achieve this function of the distribution network.

Street mains of the distribution network are fed water by the pumping station. These lines run beneath selected streets and connect to block or area distribution lines at street intersections. All mains and distribution lines must be buried to a level that will protect them from freezing and street vibration.

During normal operational periods an excessive amount of water may be demanded at various times. In some cases the pumping station may not be capable of filling the demand. To alleviate this type of problem, storage tanks or water towers are placed in the system to provide additional water. These tanks are designed to maintain system pressure at a uniform level. During peak consumption times, the water level of a storage tank may drop quite significantly. In the evening and morning hours, when water consumption is light, the storage tank is refilled.

Service connection lines distribute water from area or block lines to each individual building. The water company installs a totalizing flowmeter in the service connection line near the sidewalk or curb of a paved street. The water company owns and is responsible for all the distribution equipment up to this point. The owner of the building is responsible for all equipment beyond this connection point. As a general rule, building equipment used in the distribution of water is described as a plumbing system or simply plumbing.

BUILDING PLUMBING SYSTEMS

A plumbing system has two rather distinct parts or functions that work together in a building. The freshwater part of the system is responsible for the supply and distribution of potable water within the building. In addition to this, the system must carry dirty water and sewage out and away from the building. The sanitary drainage and vent part of the system is responsible for this function. The sewage system, as it is more commonly known, has larger pipes and must be vented into the outside air to prevent gas buildup. A composite plumbing system takes into account the operation of both system parts.

Water Distribution Plumbing

The water distribution function of a plumbing system for all practical purposes deals with equipment that is permanently installed in the building.

The owner of the building is responsible for the maintenance and upkeep of this equipment. The supply of water originates at the corporate main and is distributed throughout the building.

The location of components in a freshwater distribution system varies a great deal among different building installations. This is primarily dependent on the type of building and the intended function of each component. As a general rule, most plumbing components are designed to perform some type of control function. Valves, cocks, and faucets are used to alter the flow of water through distribution lines. Water meters measure the total amount of water that flows into the system during a given period of time. Water heaters control the temperature of water in a parallel distribution line. Each component has a specific role to play in the operation of the system. If one component does not function properly, it may alter the performance of other parts of the system.

The component layout of a freshwater distribution system is shown in Figure 7-3. Beginning at the water main this includes:

Building cutoff valve. A shutoff valve located immediately inside the building structure. It controls the flow of fresh water to the entire building.

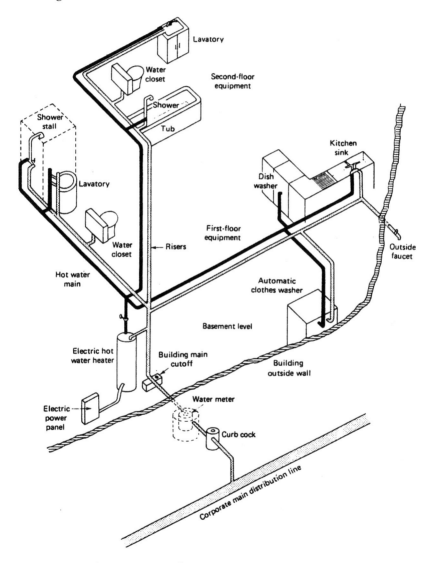

Figure 7-3. Freshwater system components.

Building main. A primary water feed line that distributes fresh water to the entire system.

Corporation cock. A cutoff valve used to control the flow of water from the corporation main to the building.

Fixture supply lines. Pipes or lines that carry water to individual fixture control valves.

Water heater. A component used to alter the temperature of water supplied by the building main.

Hot-water main. A secondary line-connected water heater outlet for distribution of hot water to selected fixtures.

Lavatories, sinks, and bathtubs. Special plumbing attachments designed to catch and hold water for washing and then let it drain away after use. They are made of porcelain or stainless steel, which will not harbor bacteria and can stand repeated scrubbing over a period of years.

Risers. Vertical pipes or lines that distribute water between floors or from one level to a higher level.

Toilets or water closets. Plumbing fixtures designed to dispose of body wastes by flushing them into the sanitary sewer system.

Water service line. A buried line of pipe that connects the main, the water meter, and the building interior.

Water meter. A water flowmeter that measures the amount of water consumed in a given period of time. A totalized meter is read from the top after the cover is removed.

Sanitary Drainage Plumbing

The sanitary drainage and vent system of a building is primarily responsible for the removal of dirty or wasteborne water. This building function may be handled collectively or privately, depending upon the design of the system. Collective sewage removal is much more efficient and sanitary than private disposal. In metropolitan areas, sewage is removed almost entirely by a collective sanitary sewer system. This system connects to each building and provides a common flow path to the area sewage treatment plant. Raw sewage is eventually converted into fertilizer and water at this plant.

Each building, in a sense, serves as a primary source of sewage for the system. Flow from the building is achieved by the force of gravity. For this to be effective, the building sewer line must be vented into the atmosphere. This admits air to the system and permits noxious gases to escape. The entire installation is made of cast iron or plastic pipe which has an inside diameter of 1-1/4 to 4 in. (3.2 to 10 cm). All horizontal lines of the system must be pitched slightly so that water will float solid materials downstream into the corporate sewer.

At each fixture of the system, drain passages are equipped with a special U- or S-shaped bend called a *trap*. The function of this trap is to retain a small amount of pipe water in such a way that it forms a seal between the drain line and the inside room air. This seal prevents gases, bacteria, and vermin from entering the building and blocks and passage of warm or cool air from the sewer line. The force of atmospheric pressure inside the drain line and outside room pressure holds water in the trap. Additional water poured into the fixture forces trap water into the drain and establishes a different seal. A new water seal is formed each time the fixture is used.

The venting or air-return function of a sanitary sewer is achieved by extending a connecting line from the drain into the open air. When a drain pipe runs full, even for a short period of time, it causes air to be forced ahead of the now and that on the return side to be siphoned away. For the system to be effective, air must be continually supplied to the drain. The main vent pipe, which extends from the drain to the roof, serves as the air source of system. Functionally, the stack vent is responsible for maintaining a water seal in traps, carrying away gases, and ventilating air into the drain to prevent lines from being coated with material deposits.

The component layout of the sanitary drain network of a building is shown in Figure 7-4. Included in the system starting at the corporation sewer line are building sewer line, main cleanout, main soil stack, stack vent, branch draws, horizontal branch lines, fixture vents, and the secondary stack. The function of each is:

Branch drains. Smaller sewer lines that connect alternate fixtures to the main soil stack. These lines are smaller, depending upon the number of fixtures being served.

Building sewer line. A carrier of raw sewage from a building into the corporate sewer line.

Fixture vents. Interconnecting lines between specific fixtures and the main soil stack for maintaining proper trap pressure.

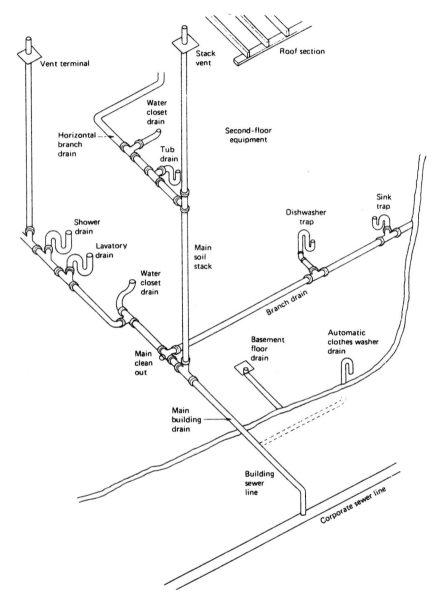

Figure 7-4. Sanitary sewer components.

Horizontal branch lines. Horizontally installed pipes that carry waste material from fixture branches to the main soil stack. Each line must be pitched downstream to produce an obstructed flow.

Main cleanout. An access connection to the main drain located near the building sewer line. This access connection permits drain cleaning between the building and the corporation sewer line.

Main soil stack. A primary flow feedline for waste material from each fixture group. This line generally extends vertically from the main sewer line to the roof stack vent.

Secondary stack. An alternate vertical line that carries waste from remote fixtures other than water closets. The upper part of the stack extends through the roof and responds as a secondary stack vent.

Stack vent. An extension of the main soil stack through the roof. Air is admitted to the system through this line and gases are expelled.

Purification

There has been a steadily growing public concern regarding the safety of the quality of water supplied by public water treatment facilities. Due to possible pollution by toxic runoff, or contaminated rivers or lakes, public concern exists regarding the quality of water. Despite treatments such as filtration and chlorination, many individuals feel that impurities exist in the potable (drinking) water supply. As a result, a market for purifying water prior to consumption has been steadily rising. For example, the growth of the bottled water industry has boomed in the past decade. Whether it is a point of use purification device such as a filter on a faucet, a commercial system for a restaurant, or a system to treat an entire commercial complex, purification systems are now becoming increasingly common place in new water distribution plumbing systems. These water filtering, or water treatment devices can utilize many different technologies to purify the water.

Filtration

Perhaps the simplest and most popular method of water purification is filtration. By using filters made of paper, activated charcoal, fiberglass, or other material that has the ability to trap small particles, water filters can remove of foreign matter from water supply systems as small as one micron. These filters can be placed in the central supply to the dwelling, such a 'whole house' filters, or placed at the point of use such as faucet filters, drinking water pitchers, or inline filters for water supplies to refrigerators, freezers, or ice machines. Aside from mechanical filtration, other methodologies exist for removal of contaminants from potable water. Distiller or boiling water filters heat the incoming water to a boiling point, the by collecting the condensate from the steam provide a distilled water output. Other systems involve running the water past magnets or electrodes to separate minerals and other deposits. The effectiveness of any water filtration device can be further enhanced by passing the water by ultraviolet (UV) light to kill the bacteria in the water.

Reverse Osmosis

Osmosis, the process of passing a liquid through a porous membrane or sheath in order to diffuse the liquid, is performed every day by living cells. Essentially, this is the premise for reverse osmosis water purification. Contaminated water is passed through a semi-permeable membrane that traps the unwanted solids in the water. When full of contaminants, the membrane is flushed and reused. Membranes have a serviceable life and must be replaced as recommended by the manufacturer of the system.

FAUCETS

A faucet is a control device that regulates the flow of water from a distribution line. In a strict sense, this device is a valve designed to conveniently open and close the water line. It has a handle or level attached to the top of the valve. Turning the handle or moving the lever opens the valve in such a way that it permits water to flow from a spout. In most installations faucets are in integral part of a plumbing fixture.

Faucet construction, which is quite varied in design. essentially provides an opening on the bottom or back side which attaches directly to the water line. Inside the faucet, the water line feeds into a chamber that is divided by a metal wall with an opening through which water can flow. In a compression type of faucet, the handle or lever operates a screw which carries a metal disk on its end. The disk is faced with a fibrous washer that bears against the opening between the two chambers when the faucet is closed. Turning the handle or lever turns the screw assembly so that the disk opens the hole between the two chambers. This action permits water to flow through the faucet. Control of water is regulated by altering the position of the faucet handle or lever. Figure 7-5 shows a representative compression type of faucet that is widely used today.

Compression faucets are easily distinguished from others because they have separate hot and cold handles to control the flow of water. In some fixtures, the output from each faucet may be mixed or expelled from a common spout or nozzle. Each handle is an independent control that alters either the hot or the cold water. As a general rule, compression faucets are somewhat more complicated in construction than others.

Compression faucets have a tendency to develop leaks after they have been placed in operation for a period of time. The washer at the valve seat generally becomes worn and will not seat properly. This condition causes water to drip from the faucet after it has been turned off. Leaking around the handle is caused by water being forced

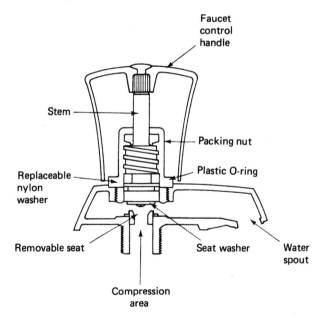

Figure 7-5. Cross-sectional view of a compression faucet.

around the rubber O-ring or packing. These two conditions can be easily corrected by replacing the O-ring or by installing a new washer. Leaky faucets cause a great deal of unnecessary waste in a water system.

Single-lever or one-knob faucets are representative of the noncompression principle of water control. These faucets achieve control without washers being pressed into a seat. In effect, noncompression faucets are classified as washerless faucets. In general, they give less trouble, and when something does occur, it is easier to repair. They are not as standardized as compression faucets, but the overall principle of control is similar enough to make repairs rather simple when the parts are available. Repair kits can be obtained at nominal cost from plumbing supply houses and hardware stores.

A wide variety of non-compression faucets are found in buildings today. These faucets are designed to mix hot and cold water and distribute it from a single spout. Control is achieved by a single knob or lever. Turning the control in a clockwise direction increases the amount of hot water delivered while reducing the cold water. Turning the knob counterclockwise increases the amount of cold water and reduces the hot water. Pulling the control knob toward the user turns it off, and pushing away increases the total flow.

The control function of a washerless faucet can be achieved by altering the position of a ball, two disks, or a cartridge. Figure 7-6 shows an example of the two-disk principle. When the control lever is lifted, it pivots the stem and lets water flow from each feedline into the spout. Pushing the control lever down stops the flow of water to the spout. Turning the handle clockwise or counterclockwise determines the mixing ratio of hot and cold water. This is achieved by altering the position of the movable disk with respect to the stationary bottom disk. The full clockwise position admits all cold water with no hot water. The extreme counterclockwise position supplies only hot water. The center or neutral position mixes equal amounts of hot and cold water.

Figure 7-7 shows an exploded view of a cartridge type of mixer faucet that is used for shower and tube water control. Pulling the control knob out or forward causes water to flow into the tub faucet or shower head. When pushed down, a diverter pin or lever on the faucet directs flow to the shower head. Water temperature is regulated

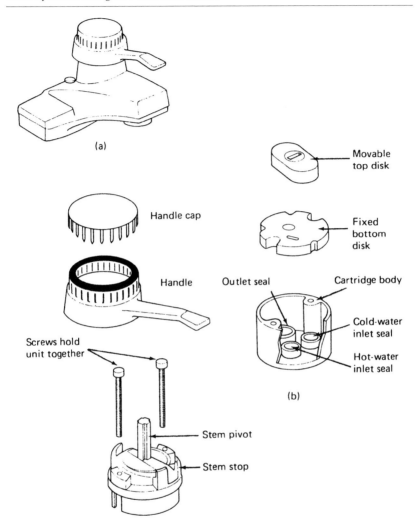

Figure 7-6. Disk-type non-compression faucet: (a) assembled; (b) internal components.

by turning the control knob clockwise or counterclockwise according to a desired temperature. Ports or holes in the cartridge regulate flow from each supply line according to its position. The entire cartridge assembly of this faucet can be replaced when leaks occur.

Ball faucets achieve the mixing function and control of water by altering the position of a ball in a socket. A kitchen sink assembly em-

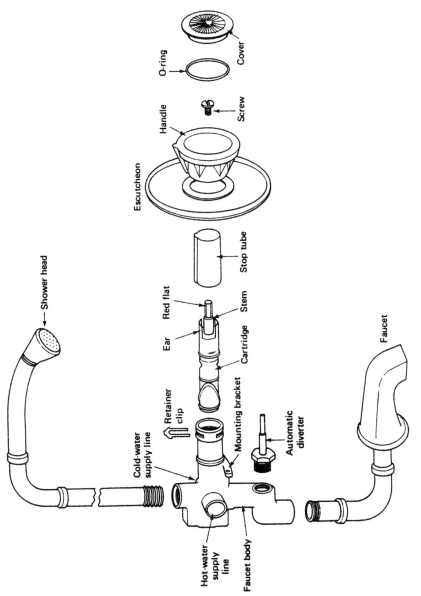

Figure 7-7. Cartridge mixer faucet assembly.

ploying the ball principle of control is shown in Figure 7-8. The operation of this faucet is very similar to the others. Thrusting the control lever back opens a flow path from each supply line through the ball to the spout. Shifting the position of the lever to the extreme right or left causes all hot or all cold water to pass into the spout. An equal mixture of hot and cold water occurs when the lever is in the center position. A diverter assembly is also available for faucets equipped with a spray assembly. When the spray lever is depressed a change in pressure forces the diverter valve to close, which causes flow to shift from spout to the spray hose. Releasing the spray valve shifts the diverter valve position so that normal spout flow is restored. The ball principle is also used in lavatory faucets and bathroom fixtures.

A relatively new development in faucets is the hands-free, no-touch, or simply automatic faucet. While designed to appear as a traditional faucet, a hands-free faucet contains an infrared device that detects when hands have been placed in front of the faucet, thereby opening a solenoid, and allowing water to pass through the faucet. Not only does this significantly reduce the amount of germs passed

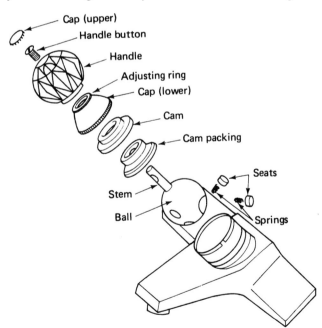

Figure 7-8. Ball-type faucet.

by coming into contact with a contaminated faucet handle, this design is very energy efficient, because it is impossible to let the water flow unless an object is present in front of the faucet. In addition to infrared, other technologies used to accomplish the same effect use photovoltaic devices that reflect ambient light from the room into the control area of the faucet to signal a user is present. Interestingly enough, this same technology is adapted to soap and disposable towel dispensers as well. While such systems have no moving parts aside from the solenoid, they do require regular maintenance in the form of maintaining a fresh battery for the infrared sensor. At least one manufacturer has designed self-sufficient system that needs no battery. It uses a small turbine in the flow of water to generate a charge, thereby eliminating the need for a battery.

Aerators

One of the easiest to install water-saving devices is the aerator. This simple device introduces air bubbles into the stream of water from a faucet, thereby restricting the flow, increasing pressure, and making it 'feel' as though water flow has increased, when in reality it has decreased by as much as 50%. Aerators generally come in two styles, those that can be installed by pushing them over the outlet of the faucet, or a more permanent mount by screwing it into a threaded faucet outlet. Aerators do not require any special skills or license to install, but must be maintained regularly by removal, inspection, and removal any foreign materials.

Toilets

Toilets, which are more precisely called water closets, are responsible for the disposal of body wastes in a building. They are made of porcelain to ensure that they will not accumulate bacteria-harboring scratches when being cleaned. With the supply water piping kept relatively small in residential buildings, most toilets are not capable of supplying the necessary force needed to completely flush waste material into the sewer line. To overcome this problem, domestic toilets are equipped with a water-storing flush tank at the back of the fixture. Water is replaced in the tank after the toilet has been flushed. The toilet is a plumbing fixture that calls for periodic maintenance in order to conserve water.

When looking into the flush tank of a toilet, one sees a number of small tubes, pipes, and unusual valves that seem rather complex. The mechanical action of the mechanism is actually rather simple when one becomes familiar with its basic operation. An understanding of fixture operation also opens the way for a number of energy conservation measures that can be applied to toilets.

Figure 7-9 shows the mechanism housed in the flush tank of a toilet. The flush handle extends outside the tank and is used to initiate the flushing action of the fixture. When pushed down, it raises the trip lever, which in turn pulls up the lift wire. The lift wire works through a guide arm that is attached to the overflow tube. Ultimately, the flush ball is lifted from the flush valve seat.

The flush ball is a hollow rubber structure made of soft rubber that normally rests on the flush valve seat. Lifting it away from the seat lets water in the flush tank flow from the tank into the toilet bowl with a great deal of force. The high-volume water flow surges into the closet bowl and washes the contents into the sewer. When this action occurs, the water level of the flush tank drops very quickly due to the large opening of the flush valve seat. The flush ball floats on the water level until it falls to the bottom of the flush valve seat. Because of this floating action, the flush handle does not need to be held down until the tank is emptied. The action is initiated with a quick twist of the lever, then a release. The flushing action stops when the flush ball reseats itself at the bottom of the tank.

With the flushing action complete, the flush tank must now be refilled. The float ball, which is made of rubber or thin metal, is designed to ride or float on the water level of the tank. Since the tank has been emptied through the flush valve, the float ball now rests near the bottom of the tank. This action causes the float valve to open, admitting water into the filler tube and bowl refill tube. As the tank begins to fill, the float rises with the increased water level. This action causes the float valve to begin its closing operation. When the float reaches its topmost position, the valve is completely closed and water flow into the tank stops. The tank has been refilled with water and the unit is prepared for the next flushing assignment. The entire flushing and filling operation should not take more than approximately 60 seconds.

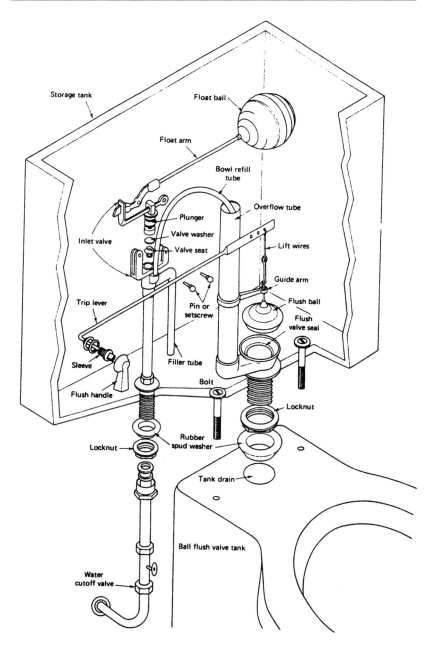

Figure 7-9. Flush-tank mechanism.

The flush tank has a supply line connected to the water main. When the float valve opens, water flows from the supply line into the tank refill tube. This action causes the tank to be filled. There is also a second line, called the bowl refill tube, connected to the float valve assembly. A small quantity of water from this line flows into the overflow tube and into the toilet bowl. This flow takes place after the flushing action has been completed. Through this supply, water returned to the toilet bowl reestablishes the water seal in the drain line. The overflow tube also serves as a guard against float valve assembly malfunction. If the valve does not shut off at the precise level, tank water will flow into the overflow tube instead of flooding the room. This condition should be corrected immediately, because it represents a flow of water into the drain. The flush tank of a domestic toilet should be checked periodically to see that it is functioning properly.

Toilets previously described use gravity to flush water from the tank to the bowl. In certain conditions, gravity does not provide ample flushing power, and pressure assisted toilets can be used. Pressure assisted toilets appear on the outside to be much the same as a conventional gravity flush toilet. The difference is in the tank on the back of the toilet. Instead of containing fill and flush assemblies as described earlier, pressure assisted toilets have another tank inside the main tank. This inner tank is sealed, and contains two chambers, one for incoming water, and the other for air. As water flows into the inner tank, it compresses the air bladder and creates pressure in the tank. When called upon to flush, the tank releases the water under the pressure from the air bladder, thereby entering the bowl at a much higher pressure than a gravity type toilet.

Government regulation has also changed the design of the toilet. In 1992, Congress passed legislation requiring all new toilets to meet a 1.6 gallon per flush (gpf) rating in order to meet the need for water conservation. Early designs of the 1.6 gpf toilets were simply traditional toilets with restrictions or dams to limit the volume of the tank. Quite often users found these toilets to be unacceptable with frequent stoppages. Changes such as drastically increasing the size of the tank drain to bowl opening, using larger trap-ways, improving the finish to the porcelain and wash-down methodologies all have improved the function of the low-flow, 1.6 gpf toilets.

Shower Heads

Shower heads are specially designed terminations that distribute water in a shower stall. Pinpoint control of water is a primary function of the shower head. Ideally, the fixture should have a swiveling capability that permits it to be adjusted to a wide range of directive positions. In addition to this, many units are equipped with a multiple-spray capability. Figure 7-10 shows some representative shower heads that can be pivoted by a swivel connection and have multispray capabilities and "low-flow" shower heads.

Flow control of a shower head can be achieved in a variety of ways. Figure 7-11a shows a comb-like structure that varies the diameter of the outlet holes in the head. With the handle up, they are wide open for a heavy wash spray. When the handle is positioned in the down position, the size of the holes are reduced for a sharp needle shower spray. Flow control is not to exceed 2.5 gallons per minute (gpm) at pressures from 15 to 120 psi. The flow control function of a shower head is an extremely important feature in building water conservation.

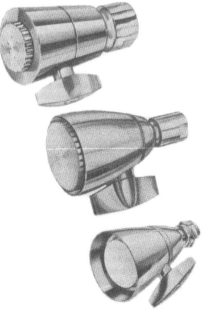

The use of small orifices or restrictors to limit flow is not recommended, rather showerheads the incorporate aeration technology are encouraged.

The Hot-water System

An ideal water system provides an adequate supply of cool, clean water to each fixture of a building. In addition to this, most buildings are equipped with an alternate source of water that has its temperature increased to a desired level. The hot-water system of a building is responsible for this function. It simply

Figure 7-10a. Typical shower heads. (*Courtesy of Chatham Brass Co., Inc.*)

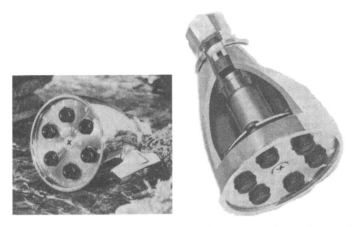

Figure 7-10b. "Low-flow" shower head. *(Courtesy of Speakman Corp.)*

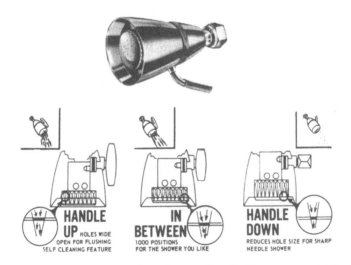

Figure 7-11. Flow control of a shower head. *(Courtesy of Chatham Brass Co., Inc.)*

receives water from the building main and increases its temperature before distribution to appropriate fixtures. Hot-water distribution lines usually run parallel and adjacent to the water system lines and are generally of the same physical size. A water heater is primarily responsible for generating all building hot water for the system.

A water heater is considered to be an essential component of the building water-supply system. Its primary function is to bring the temperature of system water up to a desired level before distributing it to each fixture. Modern water heaters are virtually maintenance-free, operate automatically, and have built-in safety devices to prevent overheating. The primary source of heat may be gas, oil, electricity, or steam, each having its own unique characteristics and peculiarities.

The selection of a water heater for a specific building application depends a great deal upon the intended function of the system. In many applications small quantities of water are needed only on a periodic basis. A number of instantaneous heaters are available to provide this type of supply. Instantaneous heaters usually provide a rather limited supply of hot water. Very small storage capacity is typical of this type of unit. Ratings for instantaneous heating units are based upon the number of gallons of hot water produced per minute. This is normally computed upon a rise in temperature of 100°F (38°C). A unit rated to produce 4 gpm of 100°F is based upon the water temperature of the main supply line. If the main supplies 50°F (10°C) water to the heating unit, it will continually deliver 150°F (65°C) at the rate of 4 gpm. If the incoming water becomes colder, the resulting output will be reduced a corresponding amount. Drawing more water from the unit than its rated capacity will also result in a lower output temperature.

Figure 7-12 shows a gas-fired copper-tube heat-exchanger water heater. This unit employs a copper-oil combustion chamber. Water circulation through the coil is heated by a gas flame from the burner. Gas fumes are exhausted through the top of the unit, which is connected into an appropriate flue. A representative unit has 160,000 Btu input per hour for recovery at 100°F (39°C) of 144 gallons. Instantaneous water heaters conserve energy because they are placed into operation only when there is a demand for hot water.

The compact water heater of Figure 7-13 is a high-capacity semi-instantaneous unit that does the work of a conventional water heater four times its size. This particular heater uses steam or boiler water as its primary source of heat. A schematic diagram of the unit is shown in Figure 7-14.

Operation of the water heater is based upon steam being applied to the inlet. It is then distributed to the steam chamber, which connects to a bundle of U-tubes. Cold water directed over the outside of

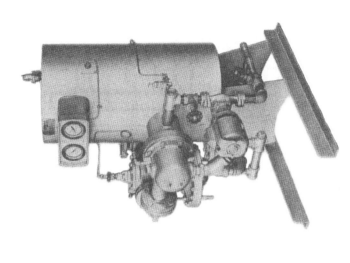

Figure 7-13. Semi-instantaneous water heater. *(Courtesy of Patterson-Kelley Co.)*

Figure 7-12. Gas-fired copper-tube water heater. *(Courtesy of A.O. Smith Company)*

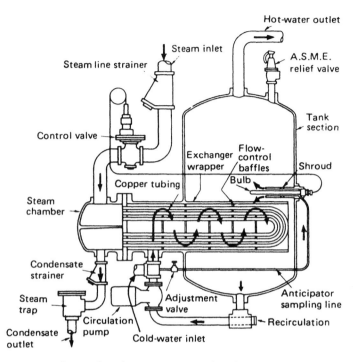

Figure 7-14. Schematic of a steam-energized water heater. (*Courtesy of Patterson-Kelley Co.*)

the U-tube bundle by the flow control baffles is heated by the time it reaches the tank section. Only a minimum volume of storage above the U-tube bundle is needed to provide an adequate source of hot water.

Figure 7-15 shows a rather new electric water heating unit designed to serve commercial and institutional buildings. This assembly is designed for reducing electrical power consumption during peak-demand periods. The heater is considerably smaller than a conventional storage unit. It holds approximately 40% more heat energy by storing water 180°F (80°C) instead of the conventional 140°F (60°C). This type of heater provides a steady flow of low input power that is regulated by the programmed controller box mounted on the front of the unit. The programmer, which is linked to the water heater element, monitors the building's electric demand and alters operation according to a predetermined demand limit. The entire assembly resets itself daily to meet varying demand changes caused by differing weather conditions.

A schematic diagram of the electric water heater of Figure 7-15 is shown in Figure 7-16. Operation is based upon cool water entering the unit at the bottom, where it is directed toward the heating element. Baffles within the heat exchanger alter the cold-water flow path so that it will be exposed to the greatest amount of heat. It then flows around the thermostatic control sensor, across the flow sensor, through a circulator, and up to the

Figure 7-15. Tank-type electric water heater. (*Courtesy of Patterson-Kelley Co.***)**

temperature-control valve, which distributes hot water to the top of the storage section. Water circulation and heating occurs only when electric power is supplied to the heating element.

Storage or tank water heaters are commonly used in residential

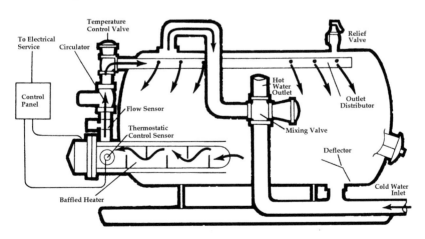

Figure 7-16. Schematic of an electric water heater. (*Courtesy of Patterson-Kelley Co.***)**

water supply systems. These units are designed to heat water and to store it until needed. Storage capacities range from 30 to 40, 50, 60, 80, and 100 gallons or more. The heating source may be gas, fuel oil, or electricity, which is placed inside of the storage tank. The selection of a specific heat source is based upon the building. Gas- and oil-heated units must be connected to a flue to exhaust unburned fumes. Electrically operated units may be located anywhere in the building structure.

In addition to the storage capacity rating of a water heater, it has a recovery-per-hour value and a Btu rating. The recovery-per-hour rating refers to the time it takes to raise the temperature of the water to a given value. A 60-gallon-per-hour recovery for 100°F (38°C) rise means that water entering the tank at 40°F (5°C) will produce 60 gallons of 140°F (60°C) water in 1 hour. The recovery rate of the heater is proportional to the size of the heating element in British thermal units. Higher Btu ratings will produce a better recovery rate.

The recovery rate of a tank water heater is directly dependent upon the type of fuel required for operation. Oil- and gas-burning units tend to have a higher recovery rate than do electric heaters of the same size. Some electric water heaters, however, are equipped with a special quick-recovery element that makes the unit compare favorably with oil and gas types. This type of unit employs an additional high-wattage element that is energized on a demand basis. Quick-recovery electric water heaters consume more electric power than do conventionally heated electric units.

All modern storage-tank water heaters have a number of common features: a storage tank, a heat source, cold- and hot-water connections, thermostatic control, and a pressure-relief valve. Figure 7-17 shows these features on three basic water heaters.

The hot-water outlet of a unit is always connected to the top of the storage tank. Hot water, being lighter, rises to the top of the tank, with cold water remaining at the bottom, where it is heated by the burner. All heating units employ thermostats to control the temperature of the tank water. Gas- and oil-burning units must be specially equipped with a pilot light that burns continuously. The pilot flame is directed on a thermocouple that regulates the main fuel valve. Should the pilot go out for some unforeseen reason, the main valve will shut down operation. Baffle plates are placed in the flue line that runs from the burner to the top of the storage tank. Diverting heat from the burner with baffles

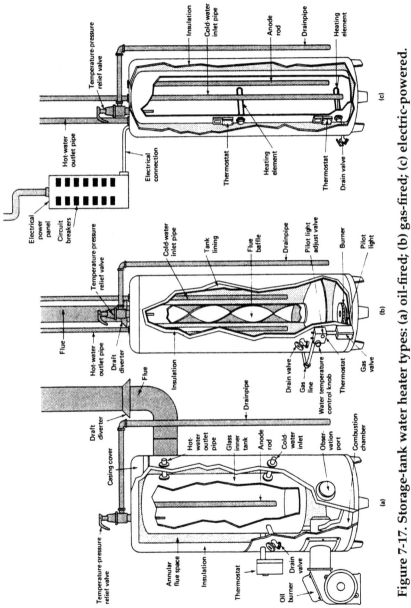

Figure 7-17. Storage-tank water heater types: (a) oil-fired; (b) gas-fired; (c) electric-powered.

helps the tank to capture interior heat very effectively.

All water heaters should be equipped with a safety device known as a relief valve. This type of valve can be actuated by either temperature or pressure, depending upon its design. Mounted at the top of the tank, the relief valve vents the storage tank if water temperature rises to a dangerous level. Should the temperature rise above 200°F (94°C) or 70 psi, the valve will open automatically. If there is no relief valve or if the existing one becomes defective, dangerously high temperatures and pres- sure can build up in the storage tank. Normal operation of a water heater causes an increase in pressure and temperature. As long as these valves remain in a normal operating range, the unit will function without relief-valve action. The relief valve is a fail-safe device that responds to equipment malfunction.

ENERGY CONSERVATION CHECKLIST
FOR WATER SYSTEMS

Energy conservation for both hot and cold water can best be achieved by reducing wasted water and making better use of any water that is being used. Further savings can be achieved by proper management of the water system. Hot-water-system management obviously has a greater potential for energy conservation than other parts of the system. Some very effective conservation measures may be implemented immediately at a nominal cost, whereas others may be classified as major cost items. A general survey of the water system usually points to a number of conservation measures that may be improved. These measures are presented in Table 7-1.

In addition to the checks and corrections listed in Table 7-1, it is a good idea to perform the following to conserve energy in water systems:

- Repair leaks including faucets and toilets.
- Use single-lever faucets.
- Use aerators.
- Install hands-free for high volume areas.
- Minimize hot water temperature.
- Purchase water heaters with high energy efficiency rating.
- Drain water heater to reduce solids such as scale.

Table 7-1: Checklist for Water Systems/Saving Our Valuable Resource

Items to check	Corrective action
Excessive water temperature levels of the hot-water system represent an unnecessary loss of energy. Test the level of hot water distributed to fixtures at several locations.	Reduce thermostat temperature setting of the water heater. Whenever possible, hot water for bathing fixtures can be reduced to 105 to 120°F (40 to 49°C). For kitchen areas and dish washing, water temperature should be 140°F (60°C).
Evaluate the insulation of hot-water storage tanks, water heaters, and hot-water pipes. Trace the domestic hot-water system through the building.	Install or replace damaged pipe or storage tank insulation. A good inch of insulation will provide a rather substantial saving in hot-water energy loss.
Investigate the operation of the electric water heater with respect to time/demand restrictions during a heating cycle.	Adjust the system to lower temperature settings for vacation periods, extended weekends, and low-use times. It may be possible to limit the duty cycle of the heater with a time clock to avoid adding the water heating load to the building load during peak electrical demand periods.
Check the hot-water system to see if it employs flow restrictors and aerators on fixtures. Hot-water utilization can be reduced approximately 50% with the installation of flow restrictors in shower nozzles and faucets.	Install flow restrictors in shower heads and fixtures that are used frequently. Replace standard faucets in restrooms with self-closing units to minimize excessive water use. Clean and check faucet aerators for mineral deposits and blockage.

Table 7-1: Checklist for Water Systems/Saving Our Valuable Resource (*Continued*)

Items to check	Corrective action
Check system fixture lines and faucets for leaks. Trace through the entire system for leaks in both the hot and cold-water distribution lines. Make a list of where leaks are evident.	Repair faucet leaks by installing new washers wherever possible. If a leak cannot be repaired, the fixture should be replaced to correct the problem. Periodically check faucets for new leaks. Hot-water faucets tend to leak more frequently than do cold-water faucets.
Evaluate the operation of building toilets or water closets. These items consume a very large portion of all building water. Check to see if water continues to flow into the water closet after it has been filled.	If water continues to flow after the closet has been filled, check the float ball and inlet valve. In some units it may be desirable to install toilet dams in the closet to conserve flushing water.
Check system water pressure. High line pressure causes banging noises when faucets are turned off and excessive water flow. Unusual vibration causes faucets and fixtures to wear out more frequently.	Measure system water pressure at several faucets and fixtures. If it is in excess of 50 psi, a pressure-reducing valve should be installed.
Investigate the combustion chamber of a gas or fuel-water heater. Look for water leaks, water accumulation in the chamber, or the smell of gas leaks.	Adjust the pilot flame to a recommended level to oil conserve fuel. Check the draft diverter and flue for dust, rust, or unnecessary obstructions. Drain water from the bottom of the tank to clear it of sediment buildup.

Chapter 8

Electrical Power Systems/ Improved Efficiency

INTRODUCTION

Electrical power is used as an energy source for heating, cooling, ventilating, lighting, and other systems used in a building. Electrical power is simple to use compared to other power systems. Usually, electrical systems require less upkeep and maintenance. This type of system also costs less to install than do other systems.

Electrical power is ordinarily purchased from a local utility company. The cost of electrical power varies from one area to another. There are many factors that determine the rate paid for electrical power that is used. These factors include the number of kilowatts or actual power usage, the demand factor or the ratio of power used to electrical power used over a period of time, and the power factor or the ratio of power in a circuit to the power delivered to a system.

An important area of energy conservation is actually knowing what and why you are conserving when studying electrical power systems. These power systems supply buildings with electrical energy for operating many types of equipment. Because of different power requirements, we must be constantly concerned with the proper operation of equipment that uses electrical power.

You will encounter various electrical and electronic symbols shown in schematic diagrams in this chapter. To assist in understanding these diagrams, a review of these symbols might be in order. A listing of these symbols follows the glossary at the end of the book.

ELECTRICAL POWER SYSTEM OVERVIEW

A block diagram of a complete electrical power system is shown in Figure 8-1. Electrical power production is an important part of the complete electrical power system. Once electrical power is produced, it must be distributed to the location where it will be used. *Electrical power distribution* systems transfer electrical power from one location to another. *Electrical power control* systems are probably the most complex of all the parts of the complete electrical power system, since there are unlimited types of devices and equipment used to control electrical power. *Electrical power conversion* systems, also called *loads,* convert electrical power into some other form of energy, such as light, heat, or mechanical energy. Conversion systems are extremely important when dealing with energy conservation. Another part of the complete electrical power system is *electrical power measurement.* Without electrical power measurement systems, control of electrical power would be almost impossible and energy conservation would be more difficult.

Each of the blocks shown in Figure 8-1 represents one important part of the complete electrical power system. For energy conservation, we should be concerned with each part of the electrical power system

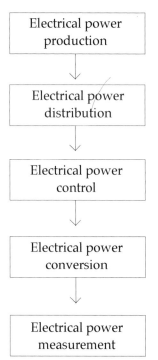

Figure 8-1. Complete electrical power system.

rather than only with isolated parts. By using the block diagram, we can develop a better understanding of how electrical power systems operate. This type of understanding is needed to help us solve energy problems that are related to electrical power consumption. We cannot consider only the production aspect of electrical power systems. We must understand and consider each part of the system, particularly the energy used by equipment.

TYPES OF ELECTRICAL CIRCUITS

There are several basic fundamentals of electrical power systems. The basics must be understood before discussing specific electrical power systems. The types of electrical circuits associated with all electrical power systems are resistive, inductive, or capacitive. Most systems have some combination of each of these three circuit types. These circuit elements are also called *loads*. A load is a part of a circuit that converts one type of energy into another type. A resistive load converts electrical energy into heat energy.

In our discussions of electrical circuits, we consider primarily alternating-current (AC) systems, as the vast majority of the electrical power produced is alternating current. Direct-current (DC) circuits are used primarily for special industrial and commercial applications.

Resistive Circuits

The simplest type of electrical circuit is a resistive circuit, such as the one shown in Figure 8-2. In resistive circuits, the following relationships exist:

$$\text{voltage } (V) = \text{current } (I) \times \text{resistance } (R)$$

$$\text{current } (I) = \frac{\text{voltage } (V)}{\text{resistance } (R)}$$

$$\text{resistance } (R) = \frac{\text{voltage } (V)}{\text{current } (I)}$$

$$\text{power } (P) = \text{voltage } (V) \times \text{current } (I)$$

These basic electrical relationships show that when voltage is increased, the current in the circuit increases in the same proportion ($V = I \times R$). Also, as resistance is increased, the current in the circuit decreases ($I = V/R$). Also, the power converted by the circuit is equal to voltage times current ($P = V \times I$). When an AC circuit contains only resistance, its behavior is similar to a direct-current (DC) circuit. Purely resistive circuits are seldom encountered in the design of electrical power systems, although some devices are primarily resistive in nature (such as resistive heating units).

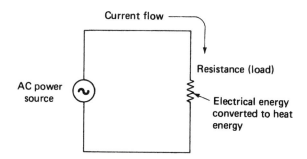

Figure 8-2. Resistive electrical circuit.

Inductive Circuits

The property of inductance is often encountered in electrical power systems. This circuit property, shown in Figure 8-3, adds more complexity to the relationship between applied voltage and current flow in an AC circuit. All motors, generators, and transformers that have wire windings exhibit the property of inductance. This property is caused by a *counterelectromotive force* (CEMF) which is produced when a magnetic force is developed around a coil of wire. The magnetic field produced around the coils affects the electrical circuit. Thus, the inductive property produced by a magnetic field offers an opposition to *change* in the current flow in a circuit.

In terms of power conversion, a *purely* inductive circuit would not convert any power in a circuit. All AC power would be delivered back to the power source. However, all circuits also have some resistance.

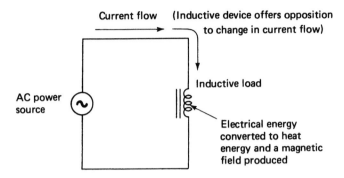

Figure 8-3. Inductive electrical circuit.

Any inductive circuit exhibits the property of inductance, which is the opposition to a change in current flow in a circuit. This property is found in coils of wire (which are sometimes called inductors), motors and generators, and transformer windings. Inductance is also present in electrical power transmission distribution lines to some extent. The unit of measurement for inductance is the *henry* (H). In an inductive circuit with AC voltage applied, an *opposition* to current flow is created by the inductance. This type of opposition is known as inductive reactance (XL).

Capacitive Circuits

Figure 8-4 shows a capacitive device connected to an AC source. When two conductive materials (plates) are separated by an insulating (dielectric) material, the property of capacitance is exhibited. De-

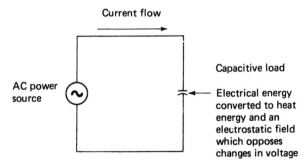

Figure 8-4. Capacitive electrical circuit.

vices called capacitors have the capability of storing electrical energy. They also have several applications in electrical power systems.

The operation of a capacitor in a circuit is dependent on its ability to charge and discharge as voltage is applied. The unit of measurement of capacitance is the *farad* (F). The farad is a very large unit, so microfarad (µF) values are ordinarily used for capacitors.

Because of the electrostatic field that is developed around a capacitor, an opposition to the flow of alternating current exists. This opposition is known as *capacitive reactance (Xc)*.

Power Relationships in AC Circuits

An understanding of basic power relationships in AC circuits is very important when studying energy conservation. In resistive circuits, power is equal to the product of voltage and current ($P = V \times I$). This formula is true only for purely resistive AC circuits. However, when a reactance (either inductive or capacitive) is present in an AC circuit, power is no longer a product of voltage and current.

Since reactive circuits cause changes in the method used to compute power, the following methods express the basic power relationships in AC circuits. The product of voltage and current is expressed in *volt-amperes* (VA) or *kilovolt-amperes* (kVA) and is known as *apparent power*. When using meters to measure power in an AC circuit, apparent power is the voltage reading multiplied by the current reading. The actual power that is converted to another form of energy by the circuit is measured with a wattmeter. This actual power is referred to as *true power*. Ordinarily, it is desirable to know the ratio of true power converted in a circuit to apparent power. This important ratio is called the *power factor* (PF) and is expressed as

$$PF = \frac{P}{VA}$$

or

$$\%PF = \frac{P}{VA \times 100}$$

where PF = power factor of the circuit
$\quad\quad\quad$ P = true power, in watts
$\quad\quad\;$ VA = apparent power, in volt-amperes

The maximum value of power factor is 1.0, or 100%, which would be obtained in a purely resistive circuit. This is referred to as unity *power factor.*

The phase angle (see Figure 8-5) between applied voltage and current flow in an AC circuit determines the power factor. In electrical circuits, the power factor varies according to the relative values of resistance and reactance (inductive or capacitive).

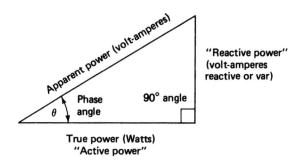

Figure 8-5. Power relationships.

The power relationships we have discussed may be simplified by looking at the power triangle shown in Figure 8-5. There are two types of power that affect the power conversion in an AC circuit. The resistive part, which results in power conversion in the circuit, is called *active power.* Active power is called the true power of the circuit and is measured in watts. The second part is that which results from an inductive or capacitive reactance and is plotted on the triangle at a 90° angle to the active power. This power, which is called *reactive power, do*es not produce an energy conversion in the circuit. Reactive power is measured in volt-amperes reactive (var).

The power triangle of Figure 8-5 shows true power (watts) on the horizontal axis, reactive power (var) at a 90° angle from the true power, and volt-amperes or apparent power (VA) as the longest side (hypotenuse) of this right triangle.

We can further examine the power relationships of the power triangle by expressing each value mathematically, based on the value of apparent power (VA) and the phase angle (θ). Remember that the phase angle is the amount of difference, in electrical degrees, between applied voltage and current now in the circuit.

Trigonometric ratios (see Appendix 2) show that the sine of an angle of a right triangle is expressed as

$$\text{sine } \theta = \frac{\text{opposite side}}{\text{hypotenuse}}$$

Since this is true, the phase angle of Figure 8-5 can be expressed as

$$\text{sine } \theta = \frac{\text{reactive power (var)}}{\text{apparent power (VA)}}$$

Therefore,

$$\text{var} = \text{VA} \times \text{sine } \theta$$

We can determine either the phase angle or the reactive power (var) value by using the trigonometric tables.

We also know that the cosine of an angle of a right triangle is expressed as

$$\text{cosine } \theta = \frac{\text{adjacent side}}{\text{hypotenuse}}$$

Thus, in terms of the power triangle,

$$\text{cosine } \theta = \frac{\text{true power (W)}}{\text{apparent power (VA)}}$$

Therefore, true power can be expressed as

$$\text{W} = \text{VA} \times \text{cosine } \theta$$

Note that the expression

$$\text{sine } \theta = \frac{\text{opposite side}}{\text{hypotenuse}}$$

is the power factor of a circuit; therefore, the power factor is equal to the cosine of the phase angle (PF = cosine θ).

Right-triangle relationships can also be expressed to determine the value of any of the sides of the power triangle when the other two values are known. These expressions are as follows:

$$VA = \sqrt{W^2 + var^2}$$

$$W = \sqrt{VA^2 - var^2}$$

$$var = \sqrt{VA^2 + W^2}$$

Appendix 2 should be reviewed to gain a better understanding of the use of right triangles and trigonometric ratios for working with electrical power problems.

ELECTRICAL POWER PRODUCTION SYSTEMS

There are millions of customers of electrical utilities companies in the United States today. To meet the demand for electrical power, power companies combine to produce vast quantities of kilowatt hours of electrical power. This electrical power is supplied by about 4000 power plants. Individual generating units which supply many megawatts of electrical power are now in operation at some power plants. These represent the *source of the electrical power system.*

Electrical power can be produced in many ways, such as from chemical reactions, heat, light, or mechanical energy. The great majority of our electrical power is produced by power plants located throughout our country which convert the energy produced by burning coal, oil, or natural gas, the falling of water, or from nuclear reactions into electrical energy. Electrical generators at these power plants are driven by steam or gas turbines or by hydraulic turbines, in the case of hydroelectric plants.

Various other methods, some of which are in the experimental stages, may be used as future power production methods. These include solar cells, geothermal systems, wind-powered systems, magnetohydrodynamic (MHD) systems, nuclear-fusion systems, and fuel cells.

Most electrical power in the United States is produced at power plants that are fossil-fuel steam plants, nuclear-fission steam plants, or hydroelectric plants. Fossil-fuel and nuclear-fission plants utilize steam turbines to deliver the mechanical energy needed to rotate the large three-phase alternators which produce massive quantities of electrical power. Hydroelectric plants ordinarily use hydraulic turbines. These units convert the force of flowing water into mechanical energy to rotate three-phase alternators.

The power plants should be located near their energy sources, near cities, or near large industries where great amounts of electrical power are consumed. The generating capacity of power plants in the United States is greater than the combined capacity of the next four leading countries of the world. Thus, we can see how dependent we are upon the efficient production and use of electrical power. We need to conserve this valuable resource.

Electrical Power Requirements

The supply and demand situation for electrical energy is much different from the other products that are produced by an organization and later sold to consumers. Electrical energy must be supplied at the same time that it is demanded by consumers. There is no simple storage system that can be used to supply additional electrical energy at peak demand times. This situation is unique and necessitates the production of sufficient quantities of electrical energy to meet the demand of the consumers at any time. Accurate forecasting of load requirements at various times must be maintained by utilities companies in order that they may predict the necessary power plant output for a particular time of the year, week, or day.

There is a significant variation in the load requirement that must be met at different times. Thus, the power plant generating capacity is subject to a continual change. For these reasons, much of the generating capacity of a power plant may be idle during low-demand times. This means that not all the generators at the plant will be in operation.

The electrical power that must be produced by our power systems varies greatly according to several factors, such as the time of the year, the time of the week, and the time of the day. Electrical power supply and demand is much more difficult to predict than most quantities that are bought and sold. Electrical power must be readily available and in sufficient quantity whenever it is required. The overall supply and demand problem is something most of us take for granted until our electrical power is interrupted. Electrical power systems in the United States must be interconnected on a regional basis so that power stations can support one another in meeting the variable load demands.

The use of electrical power is forecasted to increase every 10 years at a rate that will cause a doubling of the kilowatt hours required. Some forecasts, however, show the rate of electrical power demand to have a "leveling-off" period in the near future. This effect may be due to a saturation of the possible uses of electrical power for home appliances, industrial processes, and commercial use. These factors, combined with greater *conservation efforts,* and social and economic factors, support the idea that the electrical power demand will increase at a reduced rate in future years. The forecasting of the present demand by the electrical utilities companies must be based on an analysis by regions. The demand varies according to the type of consumer that is supplied by the power stations (which comprise the system). A different type of load is encountered when residential, industrial, and commercial systems are supplied by the utilities companies.

Industrial use of electrical power accounts for over 40% of the total kilowatt hour (kWh) consumption, and the industrial use of electrical power is projected to increase at a rate similar to its present rate in the near future.

The major increases in residential power demand have been due to increased use by customers. A smaller increase was accounted for by an increase in the number of customers. Such variables as the type of heating used, the use or nonuse of air conditioning, and the use of major appliances (freezers, dryers, ranges) affect the residential electrical power demand. At present, residential use of electrical power accounts for over 30% of the total consumption. The rate of increase will probably taper off in the near future.

Commercial use of electrical power accounts for less than 25% of the total kWh usage. Commercial power consumption includes use by office buildings, apartment complexes, school facilities, shopping establishments, and motel or hotel buildings. The prediction of the electrical power demand by these facilities is somewhat similar to the residential demand. Commercial use of electrical power is also expected to increase at a declining rate in the future.

ELECTRICAL LOAD ESTIMATING

The total electrical load of a building consists of all equipment that uses electrical energy. The amount of electrical energy used in a building depends upon the type of electrical equipment connected to the system. Typical buildings have the following types of electrical loads: (1) lighting systems; (2) HVAC systems; (3) small power systems supplied through convenience (plug-in) outlets; (4) water and sanitary systems; and (5) miscellaneous loads such as elevators, kitchen equipment, large electrical motors used to drive machines, and many other special types of electrical equipment.

In order to conserve electrical energy used in a building, it is necessary to analyze each type of electrical load attached to the electrical power system. There are many ways of reducing electrical power usage in a building. Many of these ways are discussed in this book.

As the costs of producing electrical energy continue to rise, power companies must search for ways to limit the maximum rate of energy consumption. To cut down on power usage, industries have begun to initiate programs that will cut down on the load during peak operating periods. The use of certain machines may be limited while other large power-consuming machines are operating. In larger industrial plants and at power-production plants, it would be impossible to manually control the complex regional switching systems, so computers are being used to control loads. These are discussed further in Chapter 11.

To prepare computers for power-consumption control, the peak demand patterns of local industries and the surrounding region supplied by a specific power station must be determined. The load of an industrial plant may then be balanced according to area demands with the power station output. The computer may be programmed to act as a switch, allowing only those processes to operate which are within the load calculated for the plant for a specific time period. If the load drawn by an industry exceeds the limit, the computer may deactivate part of the system. When demand is decreased in one area, the computer can cause the power system to increase power output to another part of the system. Thus, the industrial load is constantly monitored by the power company to ensure a sufficient supply of power at all times.

ELECTRICAL GENERATORS

Generators used to produce electrical power require some form of mechanical energy input. This mechanical energy is used to move electrical conductors (turns of wire) through a magnetic field inside the generator. Figure 8-6 shows the basic parts of a mechanical generator. A generator has a stationary part and a rotating part housed inside a machine assembly. The stationary part is called the *stator* and the rotating part is called the *rotor*. The generator has *magnetic field poles* of north and south polarities. Also, the generator must have a method of producing a rotary motion or a *prime mover* connected to the generator shaft. There must also be a method of electrically connecting the rotating conductors to an external circuit. This is done by a *slip ring/brush assembly.* The slip rings used on AC generators are made of copper. They are permanently mounted on the shaft of the generator. The two slip rings connect to the ends of the conductors. When a load is connected, a closed external circuit is made. With all of these generator parts functioning together, electrical power can be produced.

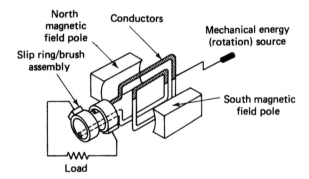

Figure 8-6. Basic parts of an electrical generator (simplified).

Single-Phase AC Generators

Although much single-phase electrical power is used, particularly in homes, very little electrical power is produced by single-phase generators. The single-phase electrical power used in homes is usually developed by three-phase generators and then converted to single-phase electricity by the power distribution system.

The current produced by a single-phase generator is in the form of what is called a sine wave. This waveform is referred to as a sine wave due to its mathematical origin, based on the trigonometric sine function. The current developed in the conductors of the generator varies as the sine of the angle of rotation between the conductors and the magnetic field. This current flow produces a voltage output. Two sine waves of single-phase AC voltage are shown in Figure 8-7.

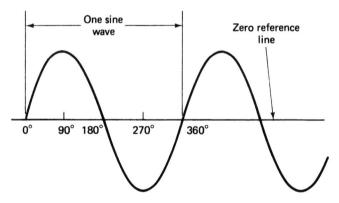

Figure 8-7. Two single-phase AC sine waves.

Three-phase AC Generators

The vast majority of electrical power produced in the United States is three-phase power. Commercial power systems use many three-phase alternators connected in parallel to supply their regional load requirements. Normally, industrial loads represent the largest portion of the load on our power systems.

A simplified diagram of a three-phase AC generator and a three-phase voltage diagram is shown in Figure 8-8. The generator output windings may be connected in either of two ways. These methods are called the *wye configuration* and the *delta configuration*. These three-phase connections are shown schematically in Figure 8-9. These methods apply not only to three-phase AC generators, but also to three-phase transformer windings and three-phase motor windings.

In the wye connection of Figure 8-9(a), the beginnings or ends of each winding are connected together. The other sides of the windings

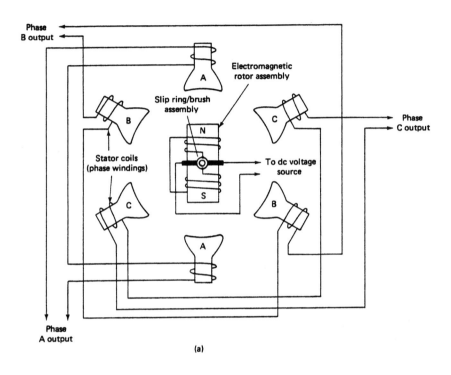

(a)

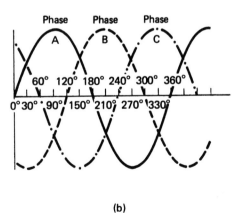

(b)

Figure 8-8. (a) Three-phase AC generator; (b) voltage diagram.

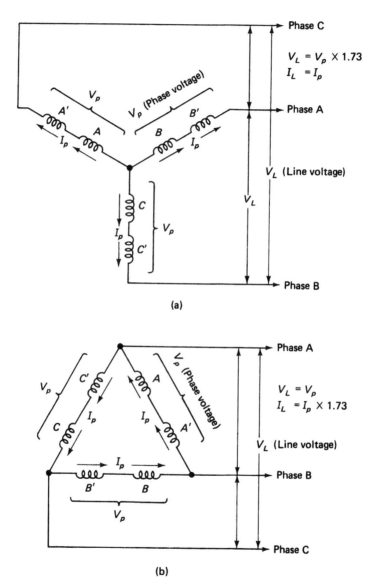

(a)

(b)

Figure 8-9. (a) Three-phase wye connection; (b) three-phase delta connection.

are the AC lines from the generator. The voltage across the AC lines (VL) is equal to the square root of 3 (1.73) multiplied by the voltage across the phase windings (Vp), or $VL = Vp \times 1.73$. The line currents (IL) are equal to the phase currents (Ip), or $IL = Ip$.

In the delta connection shown in Figure 8-9(b), the end of one phase winding is connected to the beginning to the adjacent phase winding. The line voltages (VL) are equal to the phase voltages (Vp). The line currents (IL) are equal to the phase currents (Ip) multiplied by 1.73.

The power developed in each phase (Pp) for either a wye or a delta circuit is expressed as $Pp = Vp \times Ip \times PF$, where PF is the power factor of the load. The total power (PT) developed by all three phases of a three-phase system is expressed as

$$PT = 3 \times Pp$$

$$= 3 \times Vp \times Ip \times PF$$

$$= 1.73 \times VL \times IL \times PF$$

Advantages of Three-Phase Power

Three-phase power is used primarily for industrial and commercial applications. Many types of industrial equipment use three-phase alternating-current power because the power produced by a three-phase voltage source is more constant than that produced by single-phase power sources. The effect of more power development on electric motors (with three-phase voltage applied) is that it produces a more uniform torque in the motor. This factor is very important for the large motors that are often used.

Three separate single-phase voltages can be derived from a three-phase power transmission line. It is more economical to distribute three-phase power from plants to consumers that are located a considerable distance away. Fewer conductors are required to distribute three-phase voltage. Also, the equipment that uses three-phase power is physically smaller in size than similar single-phase equipment. It is therefore more energy efficient to use three-phase power whenever possible.

ON-SITE ELECTRICAL POWER GENERATION

High utility and fuel costs have caused several large industrial and commercial organizations to consider generating their own electrical power. On-site generation provides a means of making a large energy consumer independent of local utility companies. However, the capital investment required for electrical generating equipment is great. The economics of converting a building to on-site power generation must be carefully considered. Among the factors to consider are

1. Type of prime mover used (diesel or gas engine, gas turbine, or steam turbine)

2. Type of load that is served in the building (maximum demand, variation of load, and duration of load)

3. Building occupancy schedule

4. Possibility of equipment malfunction and projected maintenance costs

5. Initial installation costs

6. Need for and availability of emergency backup power

7. Possibility of cogeneration where process steam is used

Certain buildings could economically have their own system for generating electrical power. This makes the building independent of an electrical utility company. There are several types of buildings in which electrical power could be generated economically. The overall impact of on-site power generation on energy conservation is very difficult to predict.

A major advantage of on-site power generation is that heat-recovery systems may be used to supplement the operation of the generating system. Steam or hot water from the heating system can be used for operating a turbine system which drives an on-site generator. Waste heat can be recovered to assist in the power generation process. The overall design of the HVAC system of a building and the possibility of supplementary heat recovery should be carefully analyzed to determine the effect on total energy consumption in a building.

Emergency Power Equipment

Many buildings have a need for electrical power at all times. Where electrical energy for lighting and operation of critical systems cannot be interrupted, emergency power generating systems or "standby" systems are used. Ordinarily, these systems are either battery systems or engine-driven AC generator systems. The emergency power equipment used in a building does not consume energy when not in use; however, fuel must be supplied to a generator system to assure that it will operate when needed.

When a power outage occurs, an emergency power system must be ready to operate. Usually, the system is put into operation by means of an automatic transfer switch which senses the power outage and causes the emergency system to be turned on.

DIRECT-CURRENT POWER SYSTEMS

Alternating current is used in greater quantities than is direct current; however, many important operations are dependent on direct-current power. Industries use direct-current power for many specialized processes. Electroplating and DC variable-speed motor drives are only two examples which show the need for direct-current power. Direct-current power is used to cause our automobiles to start, and for many types of portable equipment in the home. Most of the electrical power produced in the United States is three-phase alternating current, and this three-phase AC power may be easily converted to direct current for industrial or commercial use. Direct current is also available in the form of primary and secondary chemical cells. These cells are used extensively. In addition, direct-current generators are also used to supply power for specialized applications.

The conversion of chemical energy into electrical power can be accomplished by the use of electrochemical cells. Chemical cells are classified as either primary cells or secondary cells. Primary cells are ordinarily not usable after a certain period of time. After this period of time, their chemicals can no longer produce electrical energy. Secondary cells can be renewed after they are used by reactivating the chemical process that is used to produce electrical energy. This reactivation is known as charging. Both primary cells and secondary cells have many applications. When two or more cells are connected in series, they form a battery.

When a load device, such as a lamp, is connected to a battery, an electrical current will flow through the cell. Current leaves the battery through its negative terminal, flows through the load device, then reenters the battery through its positive terminal, as shown in Figure 8-10. Thus, a complete circuit is established between the battery (source) and the lamp (load), causing the lamp to light.

Regardless of the type of primary cell used, its usable time is limited. When its chemicals are expended, it becomes useless. This disadvantage is overcome by secondary cells. The chemicals of a secondary cell may be reactivated by a charging process. Secondary cells are also called storage cells.

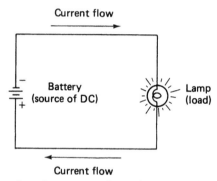

Figure 8-10. Electric current flow through a battery circuit.

POWER DISTRIBUTION SYSTEMS

Power distribution systems are an important part of our electrical systems. To transfer electrical power from a source to the place where it will be used, some type of distribution network must be utilized. Power distribution systems are used to transfer electrical power from power plants to industries, commercial buildings, and homes.

The distribution to electrical power in the United States is normally in the form of three-phase 60-Hz alternating current. Single-phase power is generally suitable for lighting and small appliances, such as those used in residential buildings. However, where a large amount of electrical power is required, three-phase power is more economical.

The distribution of electrical power involves a very complex system of interconnected power transmission lines. These transmission lines originate at the electrical power-generating stations located throughout the United States. The ultimate purpose of these power transmission and distribution systems is to supply the electrical power necessary for industrial, commercial, and residential use. From this point of view, we may say that the overall electrical power system delivers power from the source to the load that is connected to it.

Power transmission and distribution systems are used to interconnect electrical power production systems and to provide a means of delivering electrical power from the generating station to its point of utilization. Most electrical power systems east of the Rocky Mountains are interconnected with one another. These interconnections of power production systems are monitored and controlled, in most cases, by a computerized control center. Such control centers provide a means of data collection and recording, system monitoring, frequency control, and signaling. Computers have become an important means of assuring the efficient operation of electrical systems.

The transmission of electrical power requires many long interconnected power lines to carry the electrical current from where it is produced to where it is used.

Transformers

An important part of power distribution systems is an electrical device known as a transformer. Transformers are capable of controlling massive amounts of power for more efficient distribution. Transformers are also used for many other industrial and commercial applications. The distribution of alternating-current power is dependent upon the use of transformers at many points along the power distribution system. It is more economical to transmit electrical power over long distances at high voltages, since less current is required at high voltages. Substations are utilized for power distribution. At these substations, transformers are used to reduce the high transmission voltages to a lower voltage level, such as 480 volts, which is suitable for industrial motor loads, or to 120/240 volts for residential use.

Transformers provide a means of converting an alternating-current voltage from one value to another. A pictorial drawing and schematic symbol of a transformer are illustrated in Figure 8-11. Notice that the transformer shown consists of two windings which are not physically connected. The only connection between the primary and secondary windings is the magnetic coupling effect, known as *mutual induction,* which takes place when the circuit is energized by an alternating-current voltage. The laminated iron core plays an important role in transferring the magnetic field from the primary winding to the secondary winding.

If an AC current that is constantly changing in value flows in the primary winding of a transformer, the magnetic field produced around the primary will be transferred to the secondary winding. Thus, an induced voltage is developed across the secondary winding. In this way, electrical energy can be transferred from the source (primary-winding circuit) to a load (secondary-winding circuit).

Transformers are very efficient electrical devices. A typical efficiency rating for a transformer would be around 98%. Efficiency of electrical equipment is expressed as

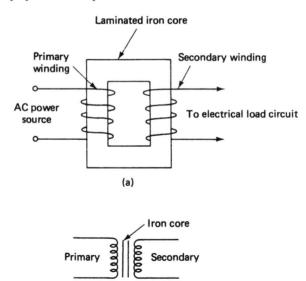

(a)

(b)

Figure 8-11. Transformer: (a) pictorial drawing; (b) schematic symbol.

$$\text{efficiency (\%)} = \frac{P_{out}}{P_{in}} \times 100$$

where

P_{out} = power output, in watts
P_{in} = power input, in watts

Transformers are classified as step-up or step-down types. These types are illustrated in Figure 8-12. The step-up transformer has fewer turns of wire on the primary than on the secondary. If the primary winding has 1 00 turns of wire and the secondary has 1000 turns, a turns ratio of 1:10 is developed. Therefore, if 10 volts AC is applied to the primary from the source, 10 times that voltage, or 100 volts ac, will be transferred to the secondary circuit. A *step-down transformer* has more turns of wire on the primary than on the secondary. If the primary winding has 500 turns while the secondary winding has 50 turns, or a 10:1 ratio, and 120 volts AC is applied to the primary from the source, then one-tenth that amount, or 12 volts ac, will be transferred to the secondary circuit.

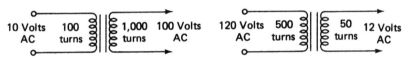

Figure 8-12. Transformer types: (a) step-up; (b) step-down.

Electrical Conductors

Conductors are used to transfer electrical current from one location to another. The unit of measurement for conductors is the *circular mil* (cmil) since most conductors are round. One mil is equal to 0.001 inch; thus, 1 circular mil is equal to a circle whose diameter is 0.001 inch. The cross-sectional area of a conductor (in circular mils) is equal to its diameter D (in mils) squared, or cmil = D2. For example, if a conductor is a 1/8 inch in diameter, its circular-mil area can be found as follows. The decimal equivalent of 1/8 inch is 0.125 inch, which equals 125 mils. Inserting this value into the formula for the cross-sectional area of a conductor gives you

$$\text{area} = D^2 \text{ (mils)}$$
$$= (125)^2$$
$$= 15{,}625 \text{ cmil}$$

If the conductor is not round in shape, its area may be found by applying the following formula:

$$\text{area (cmil)} = \frac{\text{area (square mils)}}{0.7854}$$

The resistance of a conductor expresses the amount of opposition it will offer to the flow of electrical current. The unit of measurement for resistance is the *ohm* (Ω). The resistivity (ρ) of a conductor is the resistance of a given cross-sectional area and length. This measurement is given in *circular mil-feet* (cmil-ft). The resistivity of a conductor changes with the temperature, so resistivity is usually specified at a given temperature. The resistivities for some common types of conductors are listed in Table 8-l.

Table 8-1: Resistivity of Common Conductors

Conductor	Resistivity at 20°C (W/cmil-ft)
Silver	9.8
Copper	10.4
Aluminum	17.0
Tungsten	33.0

We can use the values in Table 8-1 to calculate the resistance of any size conductor. Resistance increases as the length increases and decreases as the cross-sectional area increases. The following method can be used to find the resistance of 200 feet of aluminum conductor that is 1/8-inch in diameter. Aluminum has a resistivity of 17 ohms. The diameter (D) equals 1/8 inch, which equals 0.125 inch, which is the equivalent of 125 mils. Using the formula and substituting, we have

$$\text{resistance} = \frac{\text{resistivity} \times \text{length (feet}}{\text{diameter}^2 \text{ (mils)}}$$

$$= \frac{17 \times 200}{(125)^2} = \frac{3400}{15,625} = 0.22 \text{ ohm}$$

Conductor Sizes. Table 8-2 lists the sizes of copper and aluminum electrical conductors. The *American Wire Gauge* (AWG) is used to measure the diameter of conductors. The sizes range from No. 40 AWG, which is the smallest, to No. 0000. AWG are expressed in *million circular mil* (MCM) units. Notice in Table 8-2 that, as the AWG size number becomes smaller, the conductor is larger. Sizes up to No. 8 AWG are solid conductors, whereas larger wires have from 7 to 61 strands. Table 8-2 also lists the resistance (in ohms per 1000 feet) of copper and aluminum conductors. These values are used to determine conductor voltage drop in power distribution systems. Voltage drop has implications for energy conservation.

Ampacity of Conductors. A measure of the ability of a conductor to carry electrical current is called *ampacity*. All metal materials will conduct electrical current to some extent; however, copper and aluminum are the two most widely used. Copper is the most used metal since it is a better conductor and is physically stronger. However, aluminum is used where weight is a factor, such as for long-distance overhead power lines. The weight of copper is almost three times that of a similar volume of aluminum; however, the resistance of aluminum is over 150% that of copper. The ampacity of an aluminum conductor is, therefore, less than a similar-size copper conductor.

The ampacity of conductors depends upon several factors, such as the type of material, cross-sectional area, and the type of area in which they are installed. Conductors in the open or in "free air" dissipate heat much more rapidly than they do if they are enclosed in a metal raceway or plastic cable. When several conductors are contained within the same enclosure, heat dissipation is a greater problem.

Table 8-3 is used for conductor ampacity calculations for electrical wiring design. This table is a simplified version of those given in the National Electrical Code (NEC). The table is used to find the ampacity of conductors when not more than three are mounted in a raceway or cable.

As an example, let us find the ampacity of a No. 2 copper conductor with RHW insulation. Looking on the left column at the No. 2 wire size, we find 115 along the horizontal row under the RHW column. Thus, No. 2 copper conductors with RHW insulation will carry a maximum current of 115 Amps.

Table 8-2: Sizes of Copper and Aluminum Conductors

	Size		Number	Diameter	DC resistance at 25°C (Ω/1000 ft)	
	(AWG or MCM)	Area (cmil)	of wires	of each wire (in.)	Copper	Aluminum
AWG sizes	18	1,620	1	0.0403	6.51	10.7
	16	2,580	1	0.0508	4.10	6.72
	14	4,110	1	0.0641	2.57	4.22
	12	6,530	1	0.0808	1.62	2.66
	10	10,380	1	0.1019	1.018	1.67
	8	16,510	1	0.1285	0.6404	1.05
	6	26,240	7	0.0612	0.410	0.674
	4	41,740	7	0.0772	0.259	0.424
	3	52,620	7	0.0867	0.205	0.336
	2	66,360	7	0.0974	0.162	0.266
	1	83,690	19	0.0664	0.129	0.211
	0	105,600	19	0.0745	0.102	0.168
	00	133,100	19	0.0837	0.0811	0.133
	000	167,800	19	0.0940	0.0642	0.105
	0000	211,600	19	0.1055	0.0509	0.0836
MCM sizes	250	250,000	37	0.0822	0.0431	0.0708
	300	300,000	37	0.0900	0.0360	0.0590
	350	350,000	37	0.0973	0.0308	0.0505
	400	400,000	37	0.1040	0.0270	0.0442
	500	500,000	37	0.1162	0.0216	0.0354
	600	600,000	61	0.0992	0.0180	0.0295
	700	700,000	61	0.1071	0.0154	0.0253
	750	750,000	61	0.1109	0.0144	0.0236
	800	800,000	61	0.1145	0.0135	0.0221
	900	900,000	61	0.1215	0.0120	0.0197
	1000	1,000,000	61	0.1280	0.0108	0.0177

Table 8-3: Ampacities of Conductors in a Raceway or Cable (Three or Less)

Wire Size	Copper		Aluminum	
	With R, T, TW insulation	With RH, RHW, TH, THW insulation	With R, T, TW insulation	With RH, RHW, TH, THW insulation
14	15	15		
12	20	20	15	15
10	30	30	25	25
8	40	45	30	40
6	55	65	40	50
4	70	85	55	65
3	80	100	65	75
2	95	115	75	90
1	110	130	85	100
0	125	150	100	120
00	145	175	115	135
000	165	200	130	155
0000	195	230	155	180
250	215	255	170	205
300	240	285	190	230
350	260	310	210	250
400	280	335	225	270
500	320	380	260	310
600	355	420	285	340
700	385	460	310	375
750	400	475	320	385
800	410	490	330	395
900	435	520	355	425
1000	455	545	375	445

Insulation. Synthetic insulation for electrical wire and cable is classified into two broad categories: thermosetting and thermoplastic. The mixtures of materials within each of these categories are so varied as to make the available number of insulations almost unlimited. Most insulation is composed of compounds made of synthetic materials. These synthetic materials are combined to provide specific physical and electrical properties.

There are many types of insulation used today for electrical conductors. Some new materials have been developed that will last for exceptionally long periods of time and will withstand very high operating temperatures. The operating conditions where the conductors are used mainly determine the type of insulation required. For instance, operating voltage, heat, and moisture affect the type of insulation required. Insulation must be used that will withstand both the heat of the surrounding atmosphere and the heat developed by current flowing through the conductor. Exceptionally large currents will cause excessive heat to be in a conductor. Such heat could cause insulation to melt or burn. This is why overcurrent protection is required as a safety factor to prevent fires. The ampacity of a conductor depends upon the type of insulation used. The NEC has developed a system of abbreviations for identifying various types of insulation. Some of the abbreviations are shown in Table 8-4.

Fuses and Circuit Breakers

There are many devices that are used to protect electrical power systems from damage. For instance, switches, fuses, circuit breakers, lightning arresters, and protective relays are all used for this purpose. Some of these devices automatically disconnect electrical equipment before any damage can occur. Other devices sense changes from the normal operation of equipment and make the changes necessary to compensate for abnormal circuit conditions. The most common electrical problem that requires protection is the *short circuit*. Other problems include overvoltage, undervoltage, and changes in frequency. The purpose of any type of protective device is to cause a current-carrying conductor to become inoperative when an excessive amount of current flows through it.

The simplest type of protective device is a *fuse*. Fuses are ordinarily low-cost items and have a fast operating speed. However, in

Table 8-4: Abbreviations for Some Types of Insulation

Abbreviation	Type of insulation
R	Rubber—140° F
RH	Heat-resistant rubber—167°F
RHH	Heat-resistant rubber—194° F
RHW	Moisture and heat-resistant rubber—167° F
T	Thermoplastic—140° F
THW	Moisture and heat-resistant thermoplastic—167° F
THWN	Moisture and heat-resistant thermoplastic with nylon—194° F

three-phase systems, since each hot line must be fused, two lines are still operative if only one fuse burns out. Three-phase motors will continue to run with one phase removed. This condition is undesirable, in most instances, since motor torque is greatly reduced and overheating may result and damage the motor. Another obvious disadvantage of fuses is that replacements are required.

Circuit breakers are somewhat more sophisticated overload devices than fuses. Although their function is the same as that of fuses, circuit breakers are much more versatile. In three-phase systems, circuit breakers can open all three hot lines when an overload occurs. They may also be activated by remote-control relays. Relay systems may cause circuit breakers to open due to changes in frequency, voltage, current, or other circuit variables. The circuit breakers used in industrial plants are usually of the low voltage variety (less than 600 volts) and are housed in molded-plastic cases that are mounted in metal power distribution panels. Circuit breakers are designed so that they will automatically open when a current occurs which exceeds the rating of the breaker. Ordinarily, circuit breakers must be reset manually. Most circuit breakers employ either a thermal tripping element or a magnetic trip element.

Protective relays provide an accurate and sensitive method of protecting electrical equipment from short circuits and other abnormal conditions. Overcurrent relays are used to cause the rapid opening of electrical power lines when the current exceeds a predetermined value. The response time of the relays is very important in protecting the equipment from damage.

Motors must be protected from excessive overheating. This protection is provided by magnetic or thermal protective devices which are ordinarily within the motor-starter enclosure. Protective relays or circuit breakers can also perform this function. When an operational problem causes the motor to overheat, the protective device is used to automatically disconnect the motor from its power supply.

Power Distribution Inside Buildings

Electrical power is delivered to the location where it is to be used and then is distributed within a building by the power distribution system. Various types of circuit breakers and switch gear are employed for power distribution. Another factor involved in power distribution is the distribution of electrical energy to the many types of loads that are connected to the system. This part of the distribution system is concerned with the conductors, feeder systems, branch circuits, grounding methods, and protective and control equipment used.

Most electrical distribution to industrial and commercial loads is through wires and cables contained in raceways. Raceways hold conductors that transfer power to equipment throughout a building. Copper conductors are ordinarily used for indoor power distribution. The physical size of each conductor is dependent on the current rating of each branch circuit. Raceways may be large metal ducts or metal conduits. They provide a compact and efficient method of routing cables and wires throughout a building.

Feeder Lines and Branch Circuits

The conductors that carry current to the electrical load devices in a building are called feeders and branch circuits. Feeder lines supply power to branches that are connected to them. Primary feeder lines may be either overhead or underground. Usually, overhead lines are preferred because they permit flexibility for future expansion. Underground systems cost more, but they are much more attractive in appearance. Secondary feeders are connected to the primary feeder lines and supply power to individual sections within the building. Either aluminum or copper feeder lines may be used, depending on the specific power requirements. The distribution is from the feeder lines through individual protective equipment to branch circuits which supply the various loads. Each branch circuit has various protective devices according to the needs of the particular branch. The overall feeder-branch system may be a very complex network of switching equipment, transformers, conductors, and protective equipment.

The Electrical Service Entrance

Electrical power is brought from the overhead power lines or from the underground cable into a building by what is called a service entrance. A good working knowledge of the National Electrical Code (NEC) specifications and definitions is necessary to fully understand service-entrance equipment. The NEC sets the minimum standards that are necessary for wiring design inside a building.

The type of equipment used for an electrical service entrance of a building may include high-current conductors and insulators, disconnect switches, and protective equipment for each load circuit which will be connected to the main power system and the meters needed to measure power, voltage, current, or frequency. It is also necessary to ground the power system at the service-entrance location. This is done by a *grounding electrode,* which is a metal rod driven deep into the ground. The grounding conductor is attached securely to this grounding electrode. Then, the grounding conductor is used to make contact with all neutral conductors and safety grounds of the system.

Three-Phase Power System

Since industrial and commercial buildings generally use three-phase power, they rely upon three-phase distribution systems to supply this power. Large three-phase distribution transformers are usually located at substations adjoining the industrial plants or commercial buildings. Their purpose is to supply the proper AC voltages to meet the necessary load requirements. The AC voltages that are transmitted over long distances are high voltages which must be stepped down by three-phase transformers.

Three-phase power distribution systems that supply industrial and commercial buildings are classified according to the number of wires required. These systems, shown in Figure 8-13, are the three-phase three-wire system, the three-phase three-wire system with neutral and the three-phase four-wire system. These classifications relate to the wires that extend from the secondary windings of the power transformers. The three-phase three-wire system, shown in Figure 8-13(a), can be used to supply motor loads of 240 or 480 volts. Its major disadvantage is that it only supplies one voltage, as only three hot lines are supplied to the load. The usual insulation color code for these hot lines is black, red, or blue, as specified in the National Electrical Code.

The disadvantage of the three-phase three-wire system may be partially overcome by adding one center-tapped phase winding, as shown in the three-phase three-wire with neutral system of Figure 8-13(b). This system can be used as a supply for 120/240 volts or 240/480 volts. Assume that it is used to supply 120/240 volts. The voltage from the hot line at point 1 and the hot line at point 2 to neutral would be 120 volts, due to the center-tapped winding. However, 240 volts would still be available across any two hot lines. The neutral wire is color-coded with a white or gray insulation. The disadvantage of this system is that, when making wiring changes, it would be possible to connect a 120-volt load between the neutral and point 3 (sometimes called the "wild" phase). The voltage present here would be the combination of three-phase voltages between points 1 and 4 and points 1 and 3. This would be a voltage in excess of 200 volts. Although the "wild-phase" situation exists, this system is capable of supplying both high-power loads and low-voltage loads, such as for lighting and small equipment.

The most widely used three-phase power distribution system is the three-phase four-wire system. This system, shown in Figure 8-13(c), commonly supplies 120/208 volts or 277/480 volts for industrial or commercial load requirements. The 120/208-volt system is illustrated here. From neutral to any hot line, 120 volts for lighting and low-power loads may be obtained. Across any two hot lines, 208 volts is present for supplying motors or other high-power loads. A very popular system for industrial and commercial power distribution is the 277/408-volt system, which is capable of supplying both three-phase and single-phase loads. A 240/416-volt system is sometimes used for industrial loads, and the 120/208-volt system is also often

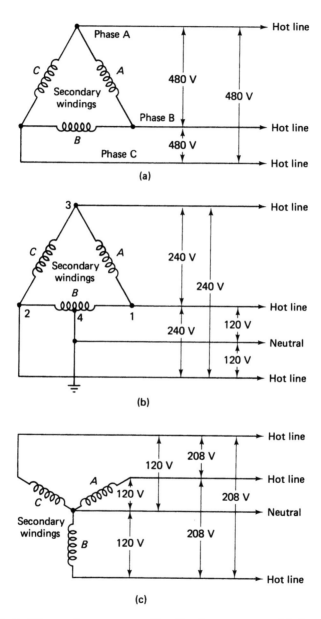

Figure 8-13. Three-phase power distribution systems: (a) three-wire; (b) three-wire with neutral; (c) four-wire.

used. Note that this system is based on the voltage characteristics of the three-phase wye connection and that the relationship $V_L = V_p \times 1.73$ exists for each application of this three-phase system.

Grounding of Distribution Systems

The concept of grounding in an electrical power system is very important. Distribution systems must have continuous uninterrupted grounds. If a grounded conductor is opened, the ground is no longer functional. An open-ground condition could present severe safety problems and cause abnormal and inefficient system operation.

Distribution systems must be grounded at substations and at the end of the power lines before the power is delivered to the load. Grounding is necessary at substations for the safety of the public and the power-company maintenance personnel. Grounding also provides points for transformer neutral connections for equipment grounds. There are two types of grounding: system grounding and equipment grounding. Another important grounding factor is ground-fault protective equipment.

System grounding involves the actual grounding of a current-carrying conductor (usually called the neutral) of a power distribution system. Three-phase systems may be either the wye or delta type. The wye system has an obvious advantage over the delta system since one side of each phase winding is connected to ground. We will define a ground as a reference point of zero-volts potential, which is usually an actual connection to earth ground. The common terminals of the wye system, when connected to ground, become the neutral conductor of the three-phase four-wire system. The delta system does not readily lend itself to grounding since it does not have a common neutral.

The second type of ground is the *equipment ground,* which, as the term implies, places operating equipment at ground potential. The conductor that is used for this purpose is either bare wire or a green insulated wire. The National Electrical Code describes conditions that require fixed electrical equipment to be grounded. Usually, all fixed electrical equipment located in industrial plants or commercial buildings should be grounded. Types of equipment that should be grounded include metal enclosures for switching and protective equipment for load control, transformer enclosures, electric motor frames, and fixed electronic test equipment. Industrial plants should use 120-volt single-phase duplex receptacles of the grounded type for all portable tools.

Ground-fault Protection

Ground-fault interrupters (GFIs) are now used extensively in industrial, commercial, and residential power distribution systems. It is required by the National Electrical Code that all 120-volt single-phase 15- or 20-ampere receptacle outlets which are installed outdoors or in bathrooms have ground-fault interrupters connected to them.

These devices are designed to eliminate electrical shock hazards resulting from individuals coming in contact with a hot AC line (line-to-ground short). The circuit interrupter is designed to sense any change in circuit conditions, such as would occur when a line-to-ground short exists. Any line-to-ground short will cause the ground-fault interrupter to open. The operating speed of the GFI is so fast, since only a very small current opens the circuit, that the shock hazard to individuals is greatly reduced.

Wiring Design

The wiring design of electrical power systems can be very complex. There are many factors that must be considered when wiring a building. Wiring design standards are specified in the National Electrical Code (NEC). The NEC, local wiring standards, and electrical inspection policies should be considered in an electrical wiring design.

When buildings are constructed or renovated, they are inspected to see if the electrical wiring meets the standards of the local ordinances, the NEC, and the local power company. The organization that supplies the electrical inspectors varies from one locality to another. Ordinarily, the local power company can be contacted for advice about electrical inspections.

Voltage Drop

Although the resistance of electrical conductors is very low, a long length of wire could cause a substantial voltage drop or reduction of voltage to the load. A voltage drop is current times resistance ($I \times R$) in the current. Whenever current flows through a circuit, a voltage drop is created. Ideally, the voltage drop caused by the resistance of a conductor will be very small.

However, a longer section of electrical conductor has a higher resistance. Therefore, it is sometimes necessary to limit the distance a conductor can extend from the power source to the load which it supplies. Many types of loads do not operate properly when a value less than the full source voltage is available. This situation can reduce the energy efficiency of equipment.

Branch Circuits

A branch circuit is defined as a circuit that extends from the last overcurrent protective device of the power system. Branch circuits, according to the National Electrical Code, are 15, 20, 30, 40, or 50 amperes in capacity. Loads larger than 50 amperes would not be connected to a branch circuit.

There are many rules in the National Electrical Code (NEC) which apply to branch-circuit design. The following information is based on the NEC. First, each circuit must be designed so that accidental short circuits or grounds do not cause damage to any part of the system. Fuses or circuit breakers are to be used as branch-circuit overcurrent protective devices. Should a short circuit or ground condition occur, the protective device will open and stop the flow of current in the branch circuit. Lighting circuits are one of the most common types of branch circuits. They are usually either 15- or 20-ampere circuits.

The maximum rating of an individual load is 80% of the branch-circuit current rating, a 20-ampere circuit could not have a single load which draws more than 16 amperes. If the load is permanently connected, its current rating cannot be more than 50% of the branch-circuit capacity if portable equipment or lights are connected to the same circuit.

Branch circuits must be designed so that sufficient voltage is supplied to all parts of the circuit. The distance that a branch circuit can extend from the voltage source or power distribution panel is therefore limited. A voltage drop of 3% is specified by the National Electrical Code as the maximum allowed for branch circuits in electrical wiring design.

A branch circuit usually consists of electrical conductors which are connected into a power distribution panel. Each branch circuit that is wired from the power distribution panel is protected by a fuse or circuit breaker. The power panel also has a main switch which controls all the branch circuits that are connected to it.

A diagram of a single-phase 120/240-volt power distribution panel is shown in Figure 8-14. Notice that eight 120-volt branch circuits and one 240-volt circuit are extended from the power panel. This type of system is used where several 120-volt branch circuits and, typically, three or four 240-volt branch circuits are required. Notice in Figure 8-14 that each hot line has a circuit breaker, while the neutral line connects directly to the branch circuits. Neutrals should never be opened (fused). This is an important safety precaution in electrical wiring.

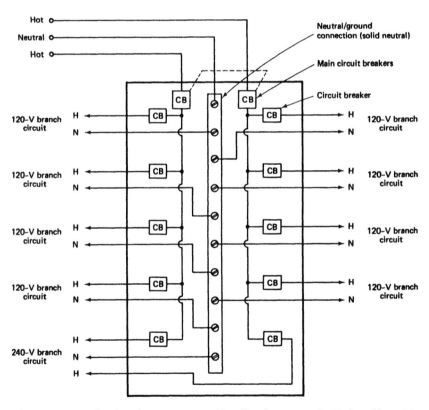

Figure 8-14. Single-phase power distribution panel: H, hot line; N, neutral line.

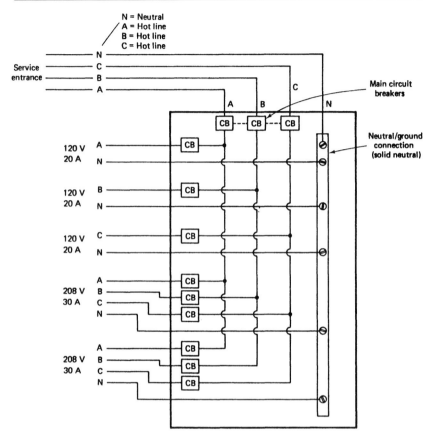

Figure 8-15. Three-phase (120/208 volt) power distribution panel: (A, hot line; B, hot line; C, hot line; N, neutral line.

A diagram for a three-phase four-wire (120/208-volt) power distribution panel is shown in Figure 8-15. There are three single-phase 1 20-volt branch circuits and two 208-volt three-phase branch circuits shown. The single-phase branches are balanced (one hot line from each branch). Each hot line has an individual circuit breaker. Three-phase lines should be connected so that an overload in the branch circuit will cause all three lines to open.

Feeder Circuits

Feeder circuits are used to distribute electrical power to power distribution panels. Many feeder circuits extend for very long distance; therefore, voltage drop must also be considered in feeder-circuit design. In higher-voltage feeder circuits, the voltage drop is reduced. However, many lower-voltage feeder circuits require large-diameter conductors to provide reduced voltage drop.

The amount of current that a feeder circuit must be designed to carry depends upon the actual load demanded by the branch-circuit power distribution panels which it supplies. Each power distribution panel will have a separate feeder circuit. Also, each feeder circuit must have its own overload protection.

Feeder-circuit design must take the conductor voltage drop into consideration. The voltage drop in a feeder circuit must be kept as low as possible so that maximum power can be delivered to the loads connected to the feeder system. The NEC allows a maximum 5% voltage drop in the combination of a branch and a feeder circuit; however, a 5% voltage reduction represents a significant energy loss in a circuit. The size of feeder conductors must be large enough to have the required ampacity, and keep the voltage drop below a specified level. If the second requirement is not met, possibly due to a long feeder circuit, the conductors should be larger than the ampacity rating required to cause the system to be more energy efficient.

ELECTRICAL POWER CONTROL

The control of electrical power systems is very important. Control is the most complex part of the electrical power system. Control equipment and devices are used in conjunction with many types of electrical loads, such as motors, lighting, and heating systems. The discussion of control systems for electrical equipment has been integrated into several other chapters in this book.

ELECTRICAL POWER CONVERSION (LOADS)

Electrical load devices that are used in buildings are very important parts of electrical power systems. The load of any system performs a function that involves power conversion. A load converts one form of energy to another form. An electrical load converts electrical energy to some other form of energy, such as heat, light, or mechanical energy. Electrical loads are usually classified according to the function they perform (lighting, heating, and mechanical). Proper operation of electrical loads is essential for energy conservation.

Load Characteristics

To plan for power system load requirements, it is necessary to understand the electrical characteristics of all the loads connected to the power system. The types of power supplies and distribution systems that a building uses are determined by the load characteristics. All electrical loads are resistive, inductive, capacitive, or a combination of these. We should be aware of the effects that various types of loads will have on the power system. The nature of alternating current causes certain electrical properties to exist.

One primary factor that affects the electrical power system is the presence of inductive loads, mainly electric motors. To counteract the inductive effects, utility companies use capacitors to correct the power factor as part of the power system design. Capacitor units are located at substations to improve the power factor of the system. The inductive effect increases the cost of a power system and reduces the actual amount of power that is converted to another form of energy. This has significant implications for energy efficiency.

Demand Factor

One electrical load relationship that is important to understand is the demand factor. The demand factor expresses the ratio between the average power requirement of a building and the peak power requirement:

$$\text{demand factor} = \frac{\text{average demand (kW)}}{\text{peak demand (kW)}}$$

The average demand for an industrial or commercial building is the average electrical power used over a specific time period. The peak demand is the maximum amount of power (kW) used during that time period. The load profile in Figure 8-16 shows a typical industrial demand versus time curve for a working day. Demand peaks that far exceed the average demand cause a decrease in the load factor ratio. Low load factors result in an additional billing charge by the utility company.

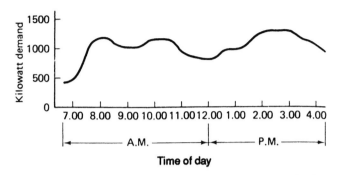

Figure 8-16. Electrical power demand curve for an industry.

Utility companies must design power distribution systems to meet the peak demand time and ensure that their generating capacity will be able to meet this peak power demand. Therefore, it is inefficient electrical design for an industry to operate at a low load factor, since this represents a significant difference between peak power demand and average power demand. Every industry should attempt to raise its load factor to the maximum level it can. By minimizing the peak demands of industrial and commercial buildings, power demand control systems and procedures can help increase the efficiency of our nation's electrical power systems.

Be careful not to confuse the load factor of a power system with the power factor. Power factor is the ratio of power converted (true power) to the power delivered to a system (apparent power).

POWER-FACTOR CORRECTION

Most industries use many electric motors; therefore, industrial plants represent highly inductive loads. This causes the power system to operate a power factor of less than unity (1.0). It is undesirable for an industry to operate at a low power factor, since the electrical power system will have to be able to supply more power to the industry than is actually used.

A fixed value of volt-amperes (voltage × current) is supplied to an industry by the electrical power system. If the power factor (PF) of the industry is low, the current is higher. The power converted by the total industrial load is equal to VA × PF. The value of power factor decreases as the reactive power (unused power) drawn by the industry increases. As reactive power is increased, more volt-amperes must be drawn from the power source. This is true since the voltage component of the supplied volt-amperes remains constant. An increase in reactive power causes a decrease in the power factor. Since

$$PF = \frac{\text{true power (watts)}}{\text{apparent power (VA)}}$$

as VA increases, the PF will decrease.

Utility companies usually charge industries for operating at power factors below a specified level. It is desirable for industries to "correct" their power factor to avoid such charges and to make more economical use of electrical energy. Two methods may be used to cause the power factor to increase: power-factor-corrective capacitors and three-phase synchronous capacitors (motors). Since the effect of capacitance is opposite to that of inductance, their effects in a circuit will counteract one another. Either power-factor-corrective capacitors or three-phase synchronous capacitors may be used to add the effect of capacitance to an AC power line.

The effect of the increased capacitive reactive power in the system is to increase or "correct" the power factor and thus reduce the current drawn from the power distribution lines that supply the loads. In many cases, it is beneficial for industries to invest in either power-factor-corrective capacitors or three-phase synchronous capacitors to correct their power factor. The advantage of synchronous capacitors over static capacitors is that their capacitive effect can be adjusted as the system power factor increases or decreases. The capacitive effect of a synchronous capacitor is easily changed by varying the DC excitation voltage applied to the rotor of the machine. Industries considering the installation of either static or synchronous capacitors should first compare the initial equipment cost and the operating cost against the savings brought about by an increased system power factor. Power-factor problems are usually not encountered in residential or commercial buildings since they do not use as many electric motors.

ELECTRICAL MOTORS

Electric motors are a very important type of electrical load. Motors convert electrical energy into mechanical energy. There are many types of motors used today. The electrical motor load is the major energy-consuming load of electrical power systems. Motors of various sizes are used for purposes that range from large industrial machine operation to the small motors that are used to power blenders and mixers in the home.

The function of a motor is to convert electrical energy into mechanical energy in the form of a rotary motion. To produce a rotary motion, a motor must have an electrical power input. Motion is produced in a motor due to the interaction of a magnetic field and a set of conductors.

All motors, regardless of whether they operate from an AC or a DC power line, have several basic characteristics in common. Their basic parts include (1) a stator, which is the frame and other stationary components; (2) a rotor, which is the rotating shaft and its associated parts; and (3) auxiliary equipment, such as a brush/commutator assembly for DC motors or a starting circuit for single-phase AC motors. The basic parts of a DC motor are shown in Figure 8-17.

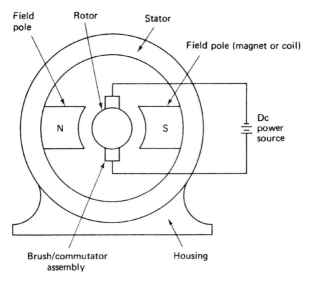

Figure 8-17. Basic parts of a DC motor.

The rotating effect of a motor is produced by the interaction of two magnetic fields and is called *torque*. The torque produced by a motor depends on the strength of the main magnetic field and the amount of current flowing through the rotor conductors. As the magnetic field strength or the current through the conductors increases, the amount of torque or rotary motion will increase also.

Direct-current Motors

Motors that operate from direct-current power sources are often used when speed control is desirable. DC motors are classified as series, shunt, or compound-wound machines, depending on the method of wiring the armature and field windings. The permanent-magnet DC motor is another type of motor that is used for certain applications.

Horsepower Rating

The horsepower rating of a motor is based on the amount of torque (rotary motion) produced at a motor's rated full load. Horsepower, which is the usual method of rating motors, can be expressed mathematically as

$$\text{hp} = \frac{2\pi \times \text{speed} \times \text{torque}}{33,000} = \frac{\text{speed} \times \text{torque}}{5252}$$

where   hp   = horsepower rating
         2π = a constant equal to 6.28

The speed of the motor is in revolutions per minute (rpm), and the torque developed by the motor is in foot-pounds.

Universal Motors

Universal motors may be powered by either AC or DC power sources. The universal motor is designed so that it will operate with either AC or DC applied. Universal motors are used mainly for portable tools and small motor-driven equipment.

AC Induction Motors

A popular type of single-phase AC motor operates on what is called the *induction principle*. Induction motors have a solid rotor which is referred to as a squirrel-cage rotor (Figure 8-18). This type of rotor has large-diameter copper conductors that are soldered at each end to a circular connecting plate. This plate actually short-circuits the individual conductors of the rotor. When current flows in the stator windings, a current is induced in the rotor. This current is developed due to *transformer action* between the stator and rotor. The stator, which has AC voltage applied, acts as a transformer primary. The rotor is similar to a transformer secondary since current is induced into it to cause the motor to operate.

The speed of an AC induction motor may be expressed as

$$\text{speed} = \frac{\text{frequency} \times 120}{\text{number of stator poles}}$$

where speed is in rpm, frequency is in hertz (cycles per second), number of stator poles is the number of windings, and 120 is a constant.

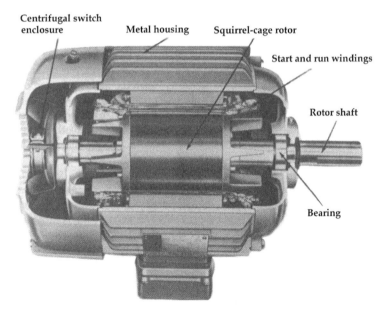

Figure 8-18. Single-phase AC induction motor—high efficiency. (*Courtesy of Westinghouse Electric Corp.*)

A two-pole motor operating at a frequency of 60 Hertz has a stator speed of 3600 revolutions per minute. The stator speed is also referred to as the synchronous speed of a motor. Actual operating speed is somewhat less.

Single-phase AC induction motors are classified according to the method used for starting. Some common types of single-phase AC induction motors include split-phase motors, capacitor motors, and shaded-pole motors.

Split-Phase Induction Motors. The split-phase AC induction motor has two sets of stator windings. One set, called the *run windings,* is connected directly across the AC line. The other set, called the *start windings,* is also connected across the AC line. However, the start winding is connected in series with a centrifugal switch that is mounted on the shaft of the motor. The centrifugal switch is in the closed position when the motor is not rotating. When the split-phase AC induction motor reaches about 75% of its normal operating speed, the centrifugal switch will open the start winding circuit since it is no longer needed. The removal of the start winding minimizes energy losses in the machine and prevents the winding from overheating. When the motor is turned off and its speed reduced, the centrifugal switch closes to connect the start winding back into the circuit and allow it to start when turned on again.

Split-phase motors are fairly inexpensive compared to other types of single-phase motors. They are used where low torque is required to drive mechanical loads such as small machinery.

Capacitor Motors. Capacitor motors are an improvement over the split-phase AC motor. Except for a capacitor placed in series with the start winding, the capacitor motor is basically the same as the split-phase motor. The purpose of the capacitor is to cause the motor to have more starting torque. The starting torque produced by a capacitor-start induction motor is much greater than that of a split-phase motor. Thus, this type of motor can be used for applications that require greater initial torque. However, they are somewhat more expensive than split-phase AC motors. Most capacitor motors, as well as split-phase motors, are used in fractional horsepower sizes (less that 1 hp). There are several types of capacitor motors used.

Shaded-Pole Induction Motors. Shaded-pole motors are used for very low torque applications such as fans and blower units. They are low-cost, rugged, and reliable motors that ordinarily come in low horsepower ratings, from 1/300 to 1/30 hp, with some exceptions. The shaded-pole motor is inexpensive since it uses a squirrel-cage rotor and has no auxiliary starting winding or centrifugal switch. Applications are limited mainly to small fans, blowers, and other low-torque applications.

Single-Phase Synchronous Motors. It is often desirable, in timing or clock applications, to use a constant-speed drive motor. Such a motor, which operates from a single-phase AC line, is called a synchronous motor. The single-phase synchronous motor has stator windings connected across the AC line. Its rotor is made of a permanent magnetic material. Once the rotor is started, it will rotate in synchronism with the applied AC voltage. The calculation of the speeds of synchronous motors is based on the following speed formula:

$$\text{speed} = \frac{\text{frequency} \times 120}{\text{number of stator poles}}$$

Therefore, for 60-Hz frequencies, the following synchronous speeds would be obtained:

Number of stator poles	Speed (rpm)
Two-pole	3600
Four-pole	1 800
Six-pole	1 200
Eight-pole	900
Ten-pole	720
Twelve-pole	600

Small synchronous motors are used in single-phase systems for low-torque applications. Such applications include clocks and timing devices that require constant speeds.

Three-Phase Induction Motors. The construction of a three-phase induction motor is very simple (Figure 8-19). It has stator windings which are connected in either a wye or a delta configuration and a squirrel-cage rotor. No external starting mechanisms are needed. Three-phase induction motors come in a variety of integral horsepower (1 hp or higher) sizes and have good starting and running torque. Three-phase induction motors are used for many applications, such as machine tools, pumps, elevators, hoists, conveyors, and other systems that use large amounts of energy.

Three-Phase Synchronous Motors. Three-phase synchronous motors are very specialized motors. They are a constant-speed motor and they can also be used to correct the power factor of three-phase systems. They are usually very large in size and horsepower rating. Physically, this motor is constructed like a three-phase alternator. Direct current is applied to the rotor and the stator windings are connected in either a wye or delta configuration. The only difference is that three-phase AC power is applied to the synchronous motor, whereas three-phase power is extracted from the alternator. Thus, the motor acts as an electrical load, while the alternator functions as a

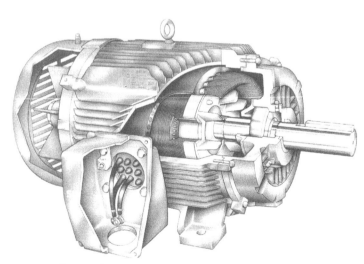

Figure 8-19. Three-phase AC induction motor—high efficiency. (*Courtesy of Siemens Energy and Automation Inc.*)

source of three-phase power.

The three-phase synchronous motor differs from the three-phase induction motor in that the rotor has windings and is connected through a slip ring/brush assembly to a DC power source.

Three-phase synchronous motors are ordinarily used in very large horsepower ratings. One method of starting a large synchronous motor is to use a smaller auxiliary DC machine connected to the shaft of the synchronous motor. Another starting method utilizes damper windings in the rotor, which are similar to the conductors of a squirrel-cage rotor. These windings are placed within the laminated iron of the rotor assembly. No auxiliary starting motor is required when damper windings are used.

A characteristic of the three-phase synchronous motor is that it can be connected to a three-phase power system to increase the overall power factor of the system. Power-factor correction was discussed previously. Three-phase synchronous motors are sometimes used only to correct the system power factor. If no load is intended to be connected to the shaft of a three-phase synchronous motor, it is called a *synchronous capacitor*. It is designed to act only as a power-factor-corrective machine. Of course, it might be beneficial to use this motor as a constant-speed drive connected to a load as well as for power-factor correction.

We know that a low power factor is undesirable in an electrical power system. Thus, the expense of installing three-phase synchronous machines could be justified by some industries to increase the system power factor. Most synchronous motors are rated greater than 100 horsepower and are used for many industrial applications requiring constant-speed drives.

Wound-Rotor Induction Motor. The wound-rotor induction motor (WRIM) is a specialized type of three-phase motor. This motor may be controlled externally by placing resistances in series with its rotor circuit. The starting torque of a WRIM motor can be varied by the value of external resistance. The advantages of this type of motor are a lower starting current, a high starting torque, smooth acceleration, and ease of control. The major disadvantage of this type of motor is that it costs a great deal more than an equivalent three-phase induction motor using a squirrel-cage rotor.

Selecting Electric Motors

Certain factors must be considered when selecting an electric motor for a specific application. Among these considerations are

1. The source voltage and power available.
2. The effect of the power factor and efficiency of the motor on the overall system.
3. The effect of the starting current of the motor on the system.
4. The type of mechanical load.
5. The expected maintenance it will require.

The major consumer of electrical energy in our country is the electric motor. It is estimated that electric motors are responsible for about 50% of the electrical power consumed. We must consider the efficient operation of motors as a major part of our energy conservation efforts. Both efficiency and power factor must be considered in determining the effect of a motor in terms of efficient power conversion. Remember the following relationships:

$$\text{efficiency (\%)} = \frac{P_{\text{out}}}{P_{\text{in}}} \times 100$$

where P_{in} is the power input, in horsepower (1 hp = 746 watts), P_{out} is the power output, in watts, and

$$\text{PF} = \frac{P}{\text{VA}}$$

where   PF = power factor of the circuit
         P = true power, in watts
        VA = apparent power, in volt-amperes

Since electrical power will probably become more expensive and less abundant, the efficiency and power factor of electric motors will become increasingly more important. The efficiency of a motor shows mathematically just how well a motor converts electrical energy into mechanical energy. A mechanical load placed on a motor affects its efficiency. Thus, it is particularly important for users to load motors so that their maximum efficiency is maintained.

Power factor is also affected by the mechanical load placed on a motor. A higher power factor means that a motor requires less current to produce a given amount of torque or mechanical energy. Lower current levels mean that less energy is being wasted (converted to heat) in the equipment and circuits connected to the motor. Penalties are assessed on industrial users by the electrical utility companies for having low system power factors (usually less than 0.8 or 0.85). By operating at higher power factors, industrial users can save money on penalties and help, on a larger scale, with more efficient utilization of electrical energy. Motor applications should be carefully studied to assure that motors (particularly very large ones) are not overloaded or underloaded, so that the available electrical energy be used more efficiently.

Voltage variation also has an effect on the power factor and efficiency. Even slight changes in voltage produce an effect on power factor. Also, an effect results when the reduced voltage causes a variation in efficiency. As proper energy utilization is becoming more and more important, motor users should make sure that their motors do not operate in undervoltage or overvoltage conditions. Proper use of motors has a great effect on energy conservation.

PRODUCTS FOR ENERGY CONSERVATION

Many electrical products have been developed recently with the purpose of energy conservation in mind. Some of these products have been discussed previously. In the following sections, some types of electrical equipment that may be used to conserve energy are discussed.

Energy-efficient Motors

The manufacturers and users of electric motors are becoming more concerned about motor efficiency. Motors consume more energy than any other type of electrical load. Therefore, the efficiency of motor operation plays a significant role in the use of electrical energy. Remember that the efficiency of electrical equipment is a comparison of energy input to energy output. For motors, efficiency is equal to power input (watts) divided by power output (horsepower). As energy costs go up, the importance of efficiency becomes more critical.

The motor shown in Figure 8-20 is manufactured by one of several companies which emphasize energy-efficient design of motors. Engineering design changes for motors have caused them to be more efficient and to operate at a higher power factor. Primarily, these design changes reduce internal energy losses, which produce heat energy rather than contributing to the production of mechanical energy. These design changes also tend to reduce motor operating temperature. The reduced temperature could increase the life expectancy of motors.

All motor applications should be studied carefully. A major problem that reduces efficiency of motors is underloading. Motor users tend to purchase motors

Figure 8-20. Magnetek "E-plus" energy-efficient motor. (*Courtesy of Magnetek, St. Louis***)**

which are larger than those actually needed for the application. A motor operated at its rated load values has the highest efficiency.

Variable-Speed-Motor Controllers

Some types of motor-driven equipment operate more economically when a variable-speed controller is used. Varying the speed of such equipment as fans and pumps can reduce their overall energy consumption. Variable-speed-motor controllers are available for direct current (DC), single-phase AC, and three-phase AC motors in a wide range of horsepower ratings. Varying the speed of AC motors is accomplished by devices called inverters, or variable frequency drive controllers (Figure 8-21). Speed of DC motors can be changed by varying input voltage. For some applications, the initial cost of an inverter or other speed-variation equipment would have a short payback period.

Automatic Cutoff Timers

Many types of electrical equipment waste energy because they are allowed to operate when they are not actually needed. The timed operation of lights was discussed in Chapter 6. The same type of timing systems can be used to control electric motor operation. Many types of timers are available to control any type of electrical load. The electromechanical timer shown in Figure 8-22 may be used to turn off a motor which controls a machine after a timed period if the machine is not in use. This eliminates having a motor operate when it is not in use or if the machine operator forgets to turn it off. In addition, an automatic stop controller could help to improve an industry's power factor since motors operating with no load have a low power factor.

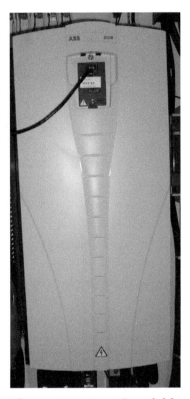

Figure 8-21. An AC variable frequency drive controller.

Single-Phase to Three-Phase Converters

Three-phase motors are generally more efficient than single-phase motors. It is desirable to use three-phase motors in as many applications as possible. Three-phase power can be developed from a single-phase power line by using a rotary converter. A rotary converter is a combination motor-generator housed in one assembly. Single-phase power causes the m-otor portion of the unit to rotate. The rotating motor is used as a prime mover to drive the three-phase generator portion of the assembly. The three-phase power output can then be used to drive three-phase equipment. Careful economic analysis can be used to determine whether the cost of a single-phase to three-phase converter could be justified for certain applications.

Figure 8-22. Automatic-stop motor controller. (*Courtesy of P.F. Industries.*)

Demand Monitors

The demand charges imposed by utility companies can often represent a significant portion of an electricity bill. A reduction in peak electrical demand can reduce the demand charge paid by a business. Peak demand of electricity is reduced by staggering the startup or operating time of large equipment. This is sometimes called *load shedding.* Electrical demand may be monitored by using a demand control relay alarm which sounds when a predetermined amount of electrical power is being used. It can also be controlled by using a profile curve developed by a chart recording instrument which monitors actual electrical energy being used over a 24-hour period. Careful analysis of peak demand times will allow a business to take corrective action in reducing demand.

ENERGY CONSERVATION CHECKLIST FOR ELECTRICAL SYSTEMS/SAVING NATURAL RESOURCES

Table 8-5 presents a checklist for electrical systems.

Table 8-5: Checklist for Electrical Systems

Items to check	Corrective action
Maintenance of electrical equipment. Check electrical equipment in the building to make sure that it is properly maintained.	Develop a preventative maintenance program for the building and enforce it; identify equipment that need, to be replaced or repaired to achieve higher energy efficiency.
Electric motors. Maintain an inventory of the location of all large electric motors used in the building; record the proper voltage, current, and horsepower ratings and assure that they are maintained.	Replace or repair motors that are not operating properly due to improper voltage or current, improper horsepower rating, loose connections, unbalanced voltage (three-phase motors), improper grounding, or unusual wear on bearings or pulleys.
Electrical control equipment. Check all electrical control equipment to assure that it operates properly.	Repair or replace control equipment that does not function properly; assure that control equipment is adjusted by authorized personnel.
Electric water heaters. Check water heaters to assure that temperature setting is as low as possible (105°F recommended) and that heaters do not operate during periods when they are not needed.	Reduce temperature setting to 105°F. Use timers to limit the duty cycle of heaters during times when they are not in full use.

Table 8-5: Checklist for Electrical Systems/Saving Natural Resources (Continued)

Items to check	Corrective action
Voltage drop in branch circuits. Check operating voltages of equipment located a long distance from the power distribution panel to assure that they do not operate at a low voltage which reduces equipment efficiency.	If possible, when equipment operates at a low voltage, replace old wire with a larger-size wire to reduce the voltage drop; in certain cases, it may be desirable to redesign the electrical power distribution system to avoid long runs of wire from distribution panels.
Electrical demand. If you pay a demand charge on your electric bill, check the building's average electrical usage (kW) and compare to the peak usage (kW) to assure that there is not a big difference. (These can usually be obtained from the electric bill or from the utility company.)	If there is a big difference in average and peak electrical usage, consider installing a demand controller, load shedder, or other timing devices which assure more uniform usage of electrical power.
Power factor of equipment. If you pay a power-factor penalty on your electric bill, check the use of large electric motors in building (this usually applies to industries only).	Check to see if motors are properly loaded during operation (underloading reduces power factor); consider installing power-factor-correcting capacitors or a three-phase synchronous capacitor motor unit to increase the power factor of the building and reduce the penalty on the electric bill.

Chapter 9

Solar Energy Systems

INTRODUCTION

Electrical power for heating and cooling must be purchased from a power company. This is the same with other fossil fuels, but power from the sun to heat and cool is free and practically limitless. On the other hand, energy from the sun is difficult to control and may be initially very expensive. Technology exists today so that we may harness this abundant energy source by using fairly inexpensive methods. Even though the price range for initial installation for a system is more expensive than one involving fossil fuels, the long-term payback is cheap. Other major advantages in using solar energy are that it can generate electricity with the use of solar cells and that it in no way harms the environment because we are simply increasing the use of this natural energy source. Even considering the problems associated with solar energy, many feel that solar energy will be the next major form of energy to be utilized extensively in the United States.

TYPES OF SOLAR ENERGY SYSTEMS

The sun delivers a constant stream of radiant energy. The amount of solar energy coming toward the earth through sunlight in one day equals the energy produced by burning millions of tons of coal. It is estimated that enough solar energy is delivered to the United States in one hour to meet the energy needs of this country for an entire year.

Solar technology can be divided into two basic types: passive and active. Passive systems use natural forces of convection and gravity to transfer heat immediately to an object or area. The other method, active, is contrasted to passive in that it makes use of mirrors to capture the sun's energy to be used at a later time.

Passive Solar Design

Passive solar design to a certain extent was used for several hundred years before the invention of electricity. This type of solar design takes advantage of the entire building in relation to the sun. Arranging a building in relation to the sun is called orientation. This also takes into account the placement of windows, quantity and quality of insulation, and basic architectural design to allow the sun to influence the heating or cooling of a building. All buildings are heated to some extent due to solar heat gain. In passive solar design, there is ordinarily no transfer fluid used to circulate heat.

There are two types of passive solar design, the direct and indirect methods. The direct method involves the transfer of solar radiation through a transparent wall into a space that requires heating. Solar energy may then be stored by interior areas that have high thermal mass, such as heavily insulated floors and walls. The heat then transfers from the area of high thermal mass to heat the interior of the building for an extended period of time. The indirect method utilizes heat gain through walls, roof ponds, and possibly through a greenhouse structure attached to a building. Roof ponds and solar greenhouses will be discussed in more detail later in the chapter.

Several factors must be considered in the passive solar design of a building. Most of these factors apply to the design of a new building. An important consideration is that solar radiation should enter from the south. A sufficient glass area should be provided along the south-facing walls. The solar energy that enters the building must then be stored and retained until heat is needed. Heat may be stored in water-filled enclosures, in concrete walls, or in floors. High-mass materials must be used to provide ample storage capacity inside of a building.

Proper shading must be provided to prevent excess heat in the summer. Shading can be accomplished by roof overhangs or awnings which are properly located. They should shade the entire glass area during the hottest time of day in the summer months. Bushy trees on the south side of a building can also be used for shading since they help block the summer sun. Dark colors that absorb sunlight can also help cool a building.

Insulated shutters should be used effectively in passive solar design. They should be closed over glass areas to prevent heat loss at night during the winter months. In the summer, the procedure can be reversed. They can be closed during the day and opened at night to provide some cooling effect.

Active Heating and Cooling Systems

An active solar heating system, such as the one outlined in Figure 9-1, consists of collector, a storage tank, and a circulating system. Ordinarily, some other type of system is used as an auxiliary or back-up system for cooler periods or times when there is prolonged cloud cover.

A major problem in solar system design is storage of the heat concentrated onto the collectors. In areas that have many cloudy days per year, an auxiliary heating system is required. As has been stated before, even considering all the problems associated with solar energy many feel it is the next major form of energy that will be used.

The system shown in Figure 9-1 is an air-type solar heating system used for space heating. The schematic diagram shows the essential parts of the system. The collector is designed so that it will capture the sun's heat and concentrate this heat into an enclosure for the air mover. They are designed with adequate insulation to minimize heat loss. The collector also has passages that allow air to flow through them, since air serves as the heat-transfer medium of the system. Any solar heating system must have collectors to absorb the sun's energy until it is used. There must be a piping system to transfer the sun's energy from the collector to the storage area.

Collector Design

The tilt angle or solar collector plates is important for efficient operation. Ordinarily, most efficient operation for space-heating and domestic hot-water systems occurs at an angle of 15 plus taking into consideration the elevation of the land that the building rests upon, or the latitude—for solar systems used for heating and cooling typically have an angle of the latitude minus 10. The orientation, in this case, of the collector plates should be exactly to the south. The size of collector required depends upon how it is used, the climatic conditions, and the

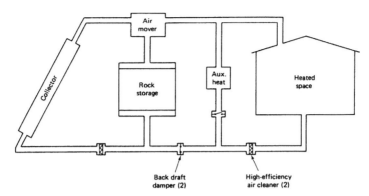

Figure 9-1. Active solar heating system. (*Courtesy of Research Products Corp.*)

type of mechanical system used with the solar system. Collectors can be located on south-facing roofs, on an adjacent building, on the ground, or on a south wall to act as an awning to shade the summer sun.

One design that is used to capture the energy of the sun makes use of parabolic, or bowl-shaped, mirrors. This design is shown in Figure 9-2. The mirrors concentrate the energy from the sun by focusing the light onto an opaque receiving surface. If tubes are provided with water circulating them, it can turn the water into steam. The steam may be used to turn a generator.

The second method of solar energy collection uses a flat-plate collector. Figure 9-3 shows some of the construction features of this particular flat-plate solar collector. Layers of glass are laid over a blackened metal plate, with an air space between each layer. The layers of glass act as a heat trap. They let the rays of the sun in but keep most of the heat from escaping. This heated air could then be used to warm a building.

Another solar collector design is called the vacuum-tube collector and is shown in Figure 9-4. Such collectors are designed to concentrate sunlight onto a glass solar absorber vacuum-tube network, solar absorber. This design makes the angle of installation less critical.

Solar Storage Systems

A common type of storage system is hot water in a metal or concrete tank. Approximately 2 to 5 gallons of storage space is needed

Figure 9-2. Solar collectors. (*Courtesy of Solar Kinetics, Inc.*)

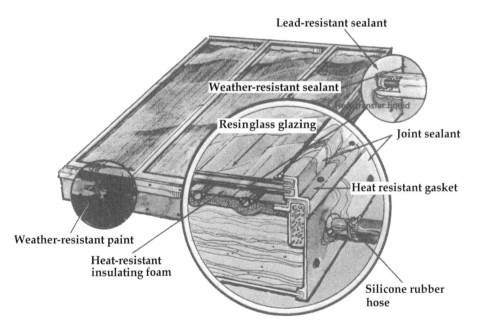

Figure 9-3. Flat-plate solar collector. (*Courtesy of Dow Corning Corp.*)

Figure 9-4. Vacuum-tube solar collector. (*Courtesy of General Electric Co., Space Division.***)**

for every square foot of the collector area. The exact amount depends upon the climate, the type of mechanical system used with the solar system, the frequency of cloudy days, and other design factors. In areas where the outdoor temperature falls below freezing, 32°F or 0°C, an antifreeze solution must be used with solar systems that use water as a storage medium. An alternative is to have an automatic system to drain the water from the collector area into the storage tanks when the temperature drops below freezing.

Figure 9-5 shows the construction of a rock storage tank which is part of an air-type solar heating system. This system provides thermal storage, which is essential since most of the heating load of buildings occurs during darkness. Thermal storage increases the overall usefulness of a solar heating system. The storage tank, shown in Figure 9-5, uses a bed of small rocks which are contained in an insulated air-tight

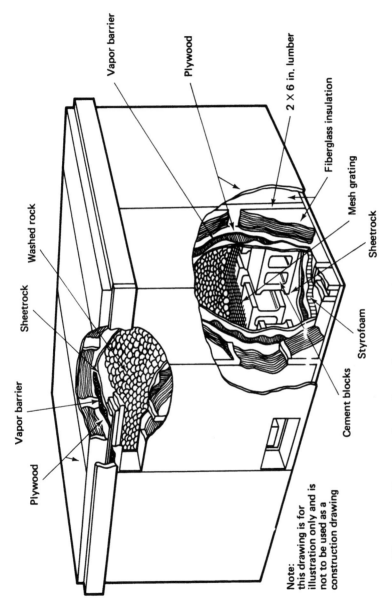

Figure 9-5. Construction of a rock storage tank. (*Courtesy of Research Products Corp.*)

enclosure. Air chambers are located above and below the rock layer so that air can be moved through the rocks. The insulation used around the storage tank retains the heat which is transferred from the collector plates and prevents heat loss into the area where the tank is located.

Air Circulation System

Another important part of a solar heating system is the air circulation system. This system uses a fan and dampers to provide air movement into the heated space and to and from the thermal storage tank. The dampers automatically control the direction of air movement accordingly to the mode of operation of the system. The air circulation system has the necessary insulated duct work to allow air movement in the desired locations.

Modes of Operation

The system shown in Figure 9-1 has four modes of operation. These operating modes are shown in Figure 9-6. The first mode, or (a), shows the heat flow directly from the collector to the heated space. When heat is needed in the building and is available plus plenty of heat is in the collectors, this mode is used. The dampers on the air circulation system are automatically set to cause air to move into the heated space. This mode of operation will continue until the space is heated to the temperature setting of the thermostat or until another mode of operation is needed.

The operating mode of Figure 9-6(b) shows the flow of the heat from the collector to the storage tank. When the temperature of the heated space is sufficient, the control system sensors determine if heat is available at the collectors to be transferred to the storage tank. One type of solar control unit is shown in Figure 9-7. The control system is used to sense temperature differences at each part of the system. During the collector-to-storage mode, the controller compares the temperature difference between the collector and the storage tank. If the difference is greater than a pre-set value, the dampers will automatically reset and allow air to circulate from the collectors to the storage tank until the temperature difference lowers again to the pre-set value.

Another operating mode is shown in Figure 9-6(c). In this mode, when little or no solar radiation is available at the collectors, heat

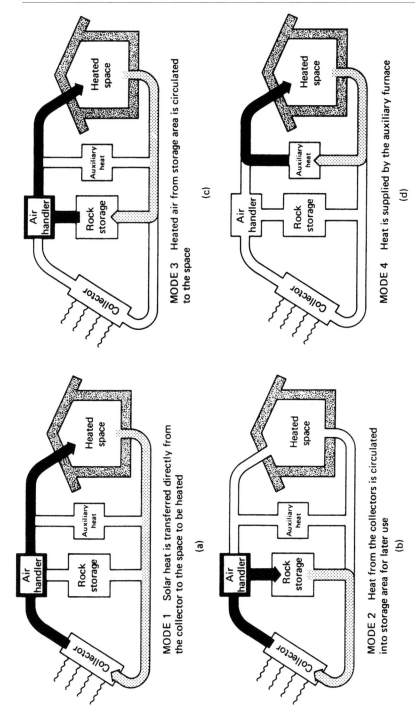

MODE 1   Solar heat is transferred directly from the collector to the space to be heated

(a)

MODE 2   Heat from the collectors is circulated into storage area for later use

(b)

MODE 3   Heated air from storage area is circulated to the space

(c)

MODE 4   Heat is supplied by the auxiliary furnace

(d)

Figure 9-6. Four operating modes of an air-type solar heating system. (*Courtesy of Research Products Corp.*)

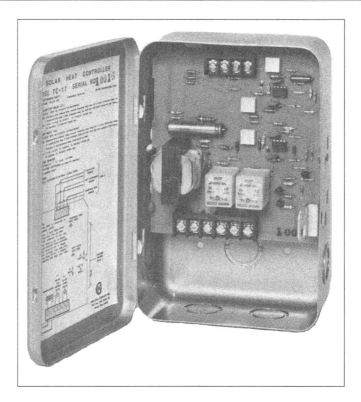

Figure 9-7. Solar-heating-system control unit. (*Courtesy of Dan-Mar Co., Inc.*)

flows directly to the storage tank to the area that needs to be heated. The dampers of the air circulation system are automatically set to allow the proper direction of the air flow. This mode of operation will continue until the space to be heated is warm enough.

The other operating mode is shown in Figure 9-6(d). The heat flow during this mode is from some type of auxiliary heating system to the heated space. When the temperatures of the collectors and the storage tank are both less than a specified minimum temperature, the dampers and associated controls are set to allow the auxiliary heating unit to circulate heated air into the building.

The layout of the air circulation duct system for an air-type solar heating system is shown in Figure 9-8. Notice in this illustration the placement of the supply and return ducts and the location of the

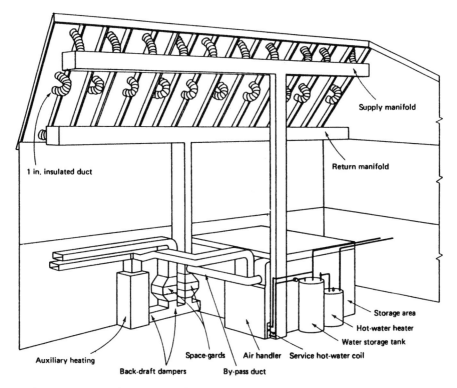

Figure 9-8. Interior layout of an air-type solar heating system. (*Courtesy of Research Products Corp.*)

dampers, auxiliary heating system, and storage area. The storage area and auxiliary heating system would normally be in the basement of a building.

SOLAR AIR-CONDITIONING SYSTEMS

Solar air-conditioning systems seem to have excellent potential for solar energy use. This is true since cooling needs are greatest when a large portion of the sun's rays reach the earth's surface. Storage requirements are less for cooling than for heating systems. But, the fluid temperature must be raised to a higher level for cooling. Solar air-conditioning systems have shown to be effective in certain geographical areas.

A solar-powered cooling unit is shown in schematic form in Figure 9-9. This system, utilized in larger buildings, is presently being developed to utilize the Rankine principle. This principle is used in a cooling system in which heat is added to a liquid to cause it to evaporate. The vapor is then expanded in an engine to produce power for rotating a compressor. The vapor from the engine is condensed and then pumped back where the cycle is repeated.

Cooling occurs through the operation of a vapor compressor air-conditioning cycle (see the right portion of Figure 9-9). The air-conditioning cycle uses a chlorofluorocarbon refrigerant, a com-

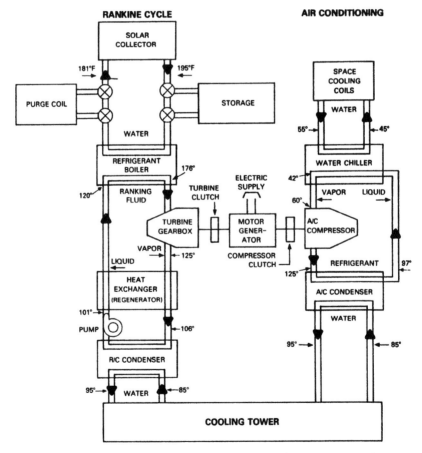

Figure 9-9. Solar-powered Rankine cooling unit. (*Courtesy of Lennox Industries, Inc.***)**

pressor, and a water-cooled condenser. A double-shaft motor-generator unit is connected to an electrical power source to supply cooling upon demand. If water in the solar unit is above 150°F (65°C), the turbine clutch is activated and the Rankine system reduces the energy required by the electric motor to operate the compressor. The cooling tower of this system serves as the heat rejecter for the air-conditioning and Rankine cycles.

PHOTOVOLTAIC SYSTEMS

Photovoltaic systems provide a method of converting solar energy directly into electrical energy. Solar, or photovoltaic, cells can be used to produce direct-current (DC) electrical energy primarily for small applications. They are used to convert light energy into electrical energy.

Photovoltaic cells were developed in the mid-1950s and were used extensively in the space program in the 1960s. Although solar cells were expensive in the beginning, their cost per watt of power has steadily decreased. Most commercial applications of photovoltaic systems are for remote communications systems, railroad switches, and for off-shore buoys. Future applications of photovoltaic systems are dependent upon many variables, including fuel costs and simplification of present manufacturing technologies.

Since the conversion that takes place in solar cells is direct energy from the sun into electrical energy, much focus has been placed on this type of system. A common application of solar cells is for photographic light-exposure meters. The electrical output of the solar cell equals the amount of light falling onto its surface.

The construction of a photovoltaic cell is shown in Figure 9-10. This cell has a layer of cadmium selenite deposited on a metal base, then a layer of cadmium oxide. In the fabrication, one layer of cadmium selenite and another layer of cadmium oxide is used. A transparent conductive film is placed over the cadmium oxide and a section of light-conductive alloy is then placed on the film. The outside leads are connected to the conductive material around the cadmium oxide layer and the metal base. When light hits the cadmium oxide layer, electrical charges are emitted and move from the (−) output.

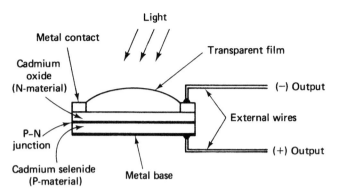

Figure 9-10a. Construction of a solar cell.

Figure 9-10b. A panel of solar photovoltaic cells.

The movement of charges caused by light energy produces a voltage between the two external terminals, so that the (+) output is from the cadmium selenite region.

The selenium solar cell has a rather low efficiency for converting light energy into electrical energy. As a rule, many cells are needed on a surface to capture the sun's radiant energy. Silicon or silicon-base cells are now being used because they are more efficient. The silicon photovoltaic cell is similar to the one shown in Figure 9-10, except that silicon is used to develop the P and N material. When light strikes the cell, a voltage is developed across the external wires. The more intense the light, the greater the voltage output.

To increase light intensity in solar cells, it has become popular to combine a cell with a light concentrator that throws more sunlight on the cell achieving greater output for a smaller cost. Because this system has a tendency to overheat, water pumped across the back of the solar cell panel collects "waste heat." Waste heat or the overabundance of heat, can be used to supply industrial, space, or water heat.

Amplification may also be used to increase the electrical output of a solar cell. This output may be used to drive a relay or some device that achieves electrical control.

The future of photovoltaics, like solar power, is hard to predict. In the near future, silicon cells will probably remain the norm with improved efficiencies and production techniques reducing their costs. Beyond that, anything is possible.

DOMESTIC SOLAR HOT-WATER HEATING

The use of solar energy to heat water is not a new idea, but there is now a renewed interest in solar hot-water heating. The parts of a solar hot-water heating system are simple. There must be a collector, a water-storage tank, a circulating pump, plus piping and control equipment. Sometimes supplementary equipment is used in more sophisticated systems. The location of the collector must be such that it will heat the water effectively during all seasons. Solar domestic hot-water systems are used mainly to supplement existing systems.

Basic System Operation

The heart of any solar hot-water heating system is the collector. The type of collector that is used with the system is shown in Figure

9-11(a), with some of its specifications listed in 9-11(b). This collector uses concentrating, nontracking collectors that are designed to produce a high-temperature (212°F) and constant temperature. An array of collectors would be used, as shown in Figure 9-12, to connect to a combination heat exchanger/storage tank. Notice that the hot water from the storage tank could be used for an absorption-type solar air-conditioning system, a solar heating system, or for solar domestic hot-water heating. In operation, sunlight enters the collector unit and is focused onto internal heat pipes by cylindrical reflectors. The concentration of heat energy inside the collector causes water in the heat pipe to vaporize and expand. Steam then flows through the piping system to the heat exchanger or water storage unit. When steam comes in contact with a surface cooler than its vaporization temperature, it will condense and release heat.

Types of Solar Hot-Water Systems

Three methods of installing solar domestic hot-water systems are shown in Figure 9-13. Figure 9-13(a) is a schematic illustration of a gravity-powered system. The collectors are mounted at a lower level than the hot-water tank. A siphoning action continually occurs which transfers energy from the solar collectors to the hot-water tank. During operation, water in the collector absorber tubes is vaporized. Steam is then transferred to a heat exchanger, where it condenses to transfer energy to water in the heat exchanger. The condenser water is then fed back to the collectors through the use of gravity to repeat the cycle.

Figure 9-13(b) shows a water-powered system. Here the collectors are mounted again at a higher level than the hot-water tank. The basic operation of this system is the same as the gravity system except condensed water from the heat exchanger is not returned to the collectors. The water can be drained through the house drainage system. The water that is vaporized in the collector absorber tubes is supplemented by tap water from the home water system.

An electricity-powered system is shown in Figure 9-13(c). The basic operation of this system is the same as the other two systems described, but the condensed water from the bottom of the steam tube is collected in the holding tank. This water is recirculated to the collectors for reuse by a small pump. Because of this recirculation, the water required for the system is reduced.

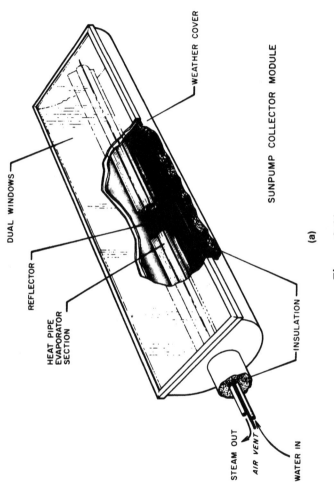

WEATHER COVER

DUAL WINDOWS

REFLECTOR

HEAT PIPE
EVAPORATOR
SECTION

INSULATION

STEAM OUT

AIR VENT

WATER IN

SUNPUMP COLLECTOR MODULE

(a)

Figure 9-11a.

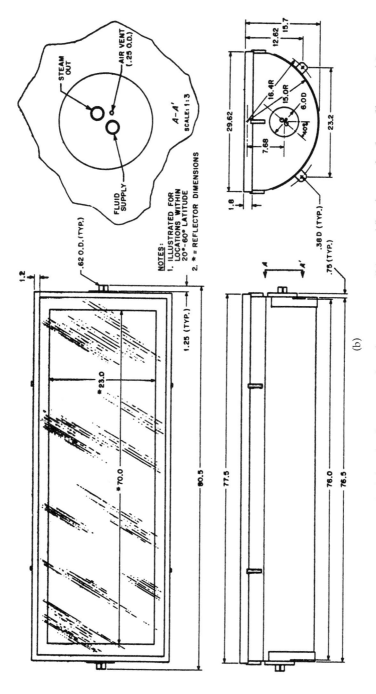

(b)

Figure 9-11. (a) Collector used with solar hot water heating system; (b) specifications of solar collector. (*Courtesy of Entropy Limited.*)

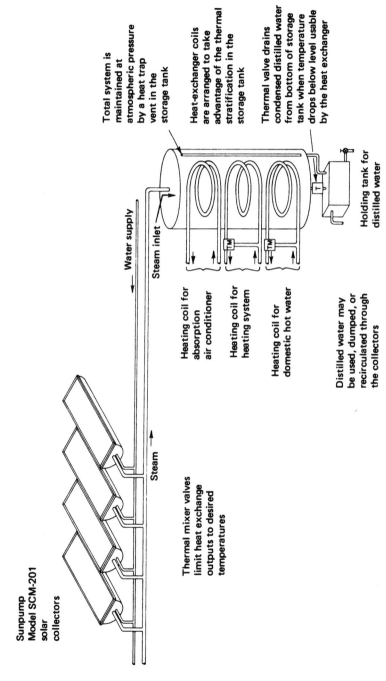

Sunpump
Model SCM-201
solar
collectors

Steam

Thermal mixer valves
limit heat exchange
outputs to desired
temperatures

Water supply

Steam inlet

Heating coil for
absorption
air conditioner

Heating coil for
heating system

Heating coil for
domestic hot water

Distilled water may
be used, dumped, or
recirculated through
the collectors

Total system is
maintained at
atmospheric pressure
by a heat trap
vent in the
storage tank

Heat-exchanger coils
are arranged to take
advantage of the thermal
stratification in the
storage tank

Thermal valve drains
condensed distilled water
from bottom of storage
tank when temperature
drops below level usable
by the heat exchanger

Holding tank for
distilled water

Figure 9-12. Solar collector array and storage/heat exchanger system. (*Courtesy of Entropy Limited.*)

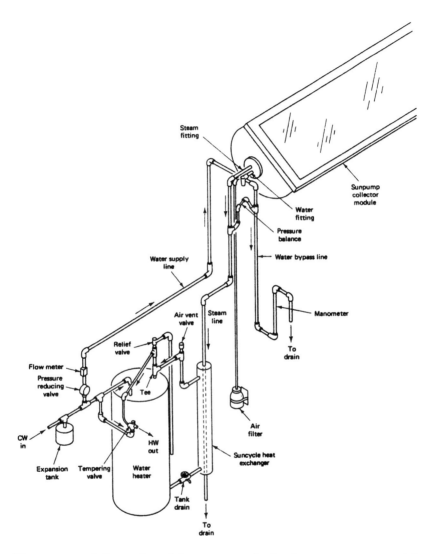

Figure 9-13a. Schematics of three types of solar hot water systems: (a) gravity-powered; (b) water-powered; (c) electricity-powered. (*Courtesy of Entropy Limited.*)

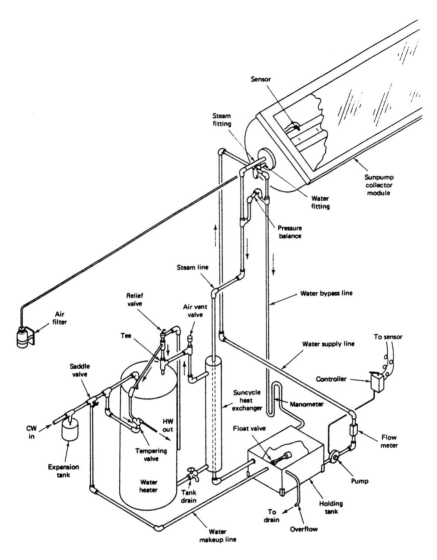

Figure 9-13b. Water-powered system.

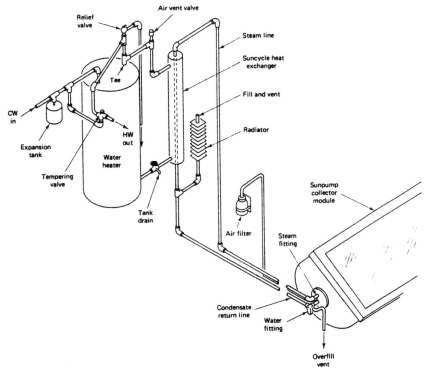

Figure 9-13c. Electricity-powered system.

PRODUCTS FOR ENERGY CONSERVATION

The sections that follow describe some products which are available and are part of solar energy systems. Keep in mind that most of these products were developed specifically for energy conservation in buildings. Solar energy systems differ from many of the other systems discussed in this book since their primary purpose is for energy conservation.

Thermal Storage Walls and Roofs

In passive solar design, thermal storage walls and roofs may be used. The concept of a thermal storage wall is illustrated in Figure 9-14. A heavy material such as concrete or water in containers are used to store solar heat. The thermal storage wall is placed directly be-

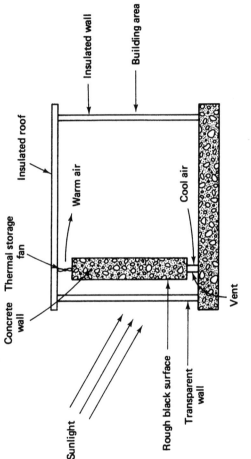

Figure 9-14. Thermal storage wall.

hind a transparent wall so that the heat produced by sunlight will be absorbed by the mass of the wall. The heat absorbed in the wall will then radiate into the building area to provide warmth during colder periods of time.

A thermal storage roof is shown in Figure 9-15. The high-mass thermal storage area of the system is placed horizontally above the ceiling of a building. The storage medium for thermal roofs is ordinarily water which is contained in an enclosure above the ceiling. Heated water will radiate through the ceiling and cause the building to be heated. When no solar radiation is available, movable insulation can be used to reduce heat loss and keep the water warmer.

Solar Greenhouses

A solar greenhouse or sunspace is designed to provide solar heat to a building. A greenhouse can be added to almost any type of building and allows a certain amount of energy conservation. Figure 9-16 shows the principle of a greenhouse. It is normally constructed on the south side of a building with heavy materials used to store the solar heat gained during the day for nighttime use. The use of a sun-

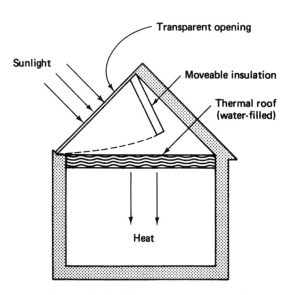

Figure 9-15. Thermal storage roof.

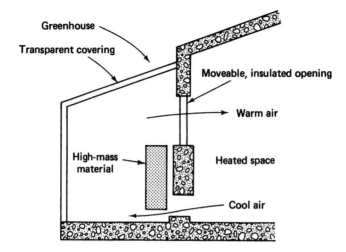

Figure 9-16. Solar greenhouse.

space such as a porch or atrium can be an attractive design feature of a building as well.

Skylights

Skylights have been used for several years to utilize sunlight or natural light for a building. Their potential for energy conservation should be considered. Skylights should be installed when natural lighting can be provided without increasing the heating and cooling costs of a building. By using proper installation and design, energy efficient skylights can be used to provide natural lighting and reduce the heating, air-conditioning, and electrical lighting requirements for a building.

Solar Roof Ponds

A solar roof pond is bags of liquid, such as water, located underneath moveable insulated panels. These panels are on tracks which slide back and forth to either cover or expose these bags. The bags are exposed during cool times of the year to gather sunlight during the day. When night arrives, they are covered by the insulating panels which re-radiate the heat back down into the building. During the daylight hours, the bags are covered with the insulating panels and at night are uncovered to discharge the heat.

Solar Swimming Pool Heaters

Active solar heating systems can be used with existing pools or newly constructed pools to heat water. Collectors for the system can be mounted on a nearby roof, a hillside, or on a frame near the pool. The operation of a swimming pool heater is the same as that of any active solar heating system. Water is heated and pumped by circulating pumps into the pool.

FUTURE OF SOLAR ENERGY

In the introduction, a somewhat rosy picture was painted of solar energy. Utilizing solar energy is definitely free, clean, safe, abundant, renewable, and the long-term payback is certainly cheap. There are some major drawbacks that have held back this ideal energy source. Primary concern plays on the initial expense of installing a solar efficient system in a building whether it be to supplement an existing system or constructing a new building. To individual contractors, the investment is a large one to construct a building. The government has also deemed that all systems be energy efficient to a certain extent. This tends to guide the public to buy the economically cheap electrical systems on the market. There is also less research to establish less expensive solar products in the marketplace.

ENERGY CONSERVATION CHECKLIST FOR SOLAR SYSTEMS/GREEN POSSIBILITY

Table 9-1 presents a checklist for solar systems.

Table 9-1: Checklist for Solar Systems

Items to check	Corrective action
Utilization of sunlight. Check to see if blinds or curtains are used on open areas to use sunlight properly.	Keep blinds and curtains closed when an area is unoccupied to reduce heat loss in winter and solar heat gain in summer; replace damaged blinds or curtains with insulated ones.
Natural lighting. Check to see if natural lighting can be used to reduce the need for electrical lighting.	Open curtains or shades to obtain more natural lighting when this is not detrimental to heat loss or gain through large open areas.
Shading. Check to see if it would be beneficial to shade certain areas of the building (exterior and interior) and reduce heat gain in summer and wind velocity in winter.	Place trees properly around the building; install reflective solar film over glass areas to reduce solar heat gain by "interior shading."
Active solar potential. Check to see if it would be cost effective to install solar heating, air conditioning, domestic hot water, or swimming pool heating systems.	Install any system that will be cost effective in your location.

Chapter 10
Instrumentation and Measurement

INTRODUCTION

A measuring instrument is simply a device used to determine the value of a quantity, a condition of operation, a physical variable or some other phenomenon. In building systems, numerous instruments are called upon to test system performance, evaluate operation, and determine efficiency. Several of these instruments will be mentioned in this chapter. In certain buildings, only about half of these mentioned will pertain. In other buildings, possibly all of these might be used. A person working with this equipment must have some understanding of the instruments that are available and be aware of common measuring techniques. The person must also be able to evaluate system performance from measured values.

Instruments used with building systems are designed primarily to display certain operating conditions within the system. In some applications instruments may be permanently attached to a system component, whereas in others it may be used to perform temporary measurements. System operation is usually dependent in some way on permanently attached instruments. Test instruments are temporarily attached to system components to evaluate specific operating conditions and to make adjustments. As a general rule, most instruments add to the total load of the system to some extent because they require a certain amount of energy to operate.

Building systems in general have something to do with the comfort of its occupants or control of operational processes. Specifically, this involves the manipulation of certain system variables. In practice, several variables may be altered in order to maintain the inside environment of a building at a desired level of comfort. Some representative variables that are subject to change in building system operation

355

are temperature, humidity, pressure, light level, airflow, and electrical power. These variables must be measured and evaluated periodically to maintain the system at its peak level of efficiency. Instrumentation and measurement are designed to perform this function.

TEMPERATURE-MEASURING INSTRUMENTS

Temperature is one of the most common and most important measurements that we make in the evaluation of building equipment today. Comfort heating equipment, air-conditioning-system operation, hot-water equipment evaluation, lighting humidity, air infiltration, and electrical equipment operation all call for some type of temperature evaluation. Precise control of temperature is a key factor in the operation of nearly any building. A person working with this equipment must be capable of making accurate temperature measurements and selecting the most appropriate instrument suited for a particular job.

Temperature-measuring instrumentation is commonly divided into two rather general classifications of electrical -or electronic and nonelectrical. Nonelectrical instrumentation was developed first and finds a rather wide range of uses in the measurement and evaluation of building equipment. It is generally easy to use, rather reliable, and has a rather simplified principle of operation. Electronic instrumentation is somewhat more complicated in operation and either responds to changes in electrical energy or utilizes electricity to produce an indicated output.

NONELECTRICAL INSTRUMENTATION

Nonelectrical temperature instrumentation responds in general to the principle of linear expansion. Changes in temperature will, in effect, cause a very pronounced change in the physical dimensions of a solid material or the expansion of liquid or gas pressure. As a result of this action, temperature values can be transformed into physical or mechanical changes that can be read on a calibrated scale.

Filled System Indicators

Filled system indicators are probably the oldest and most commonly used method of determining temperature today. The filled glass tube thermometer of Figure 10-1 measures temperature through the uniform expansion of mercury or alcohol in a sealed glass tube. The glass stem has a uniform hole bored lengthwise with a bulb reservoir at one end. When the reservoir is filled with liquid, air removed from the remainder of the tube, and the top of the tube is scaled, we have a filled thermal expansion instrument.

The glass of a filled thermometer does not expand or contract as quickly as mercury or alcohol when there is a change in temperature. This means that the enclosed liquid will rise and fall in the tube according to changes in temperature. The volumetric expansion of pure mercury is 0.01% per degree Fahrenheit and is very linear from its freezing point of –38°F (–38.8°C) to its boiling point of 1000°F (537.7°C). Calibration marks on the glass tube are used to indicate temperature on a specific scale.

The two most common filled thermometer scales in use today and the Fahrenheit (F) and Celsius (C) scales of Figure 10-1. These

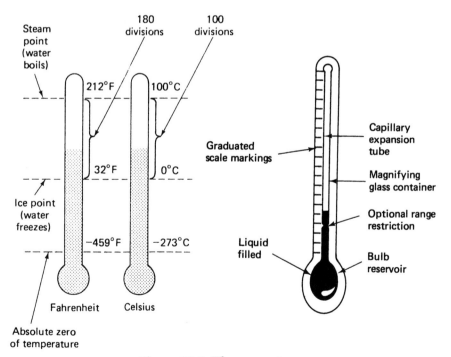

Figure 10-1. Thermometers.

scales have been established by selecting two fixed calibration points corresponding to the boiling point and ice point of water at an atmospheric pressure of 1. The assignment of scale numbers and graduated intervals is based upon a selected number of divisions. The Fahrenheit scale is divided into 180 equal divisions between the two calibration points, with the ice point 32 divisions above the zero point. The Celsius scale has 100 divisions between the two calibration points, with zero representing the ice point and 100 the boiling point. In the International System of Units (SI) adopted in 1964, the term "Celsius" is now used for degree designations that were formally called centigrade values.

Liquid-filled glass tube thermometers are widely used today to make direct temperature measurements of an operating system. In its regular form, such a thermometer is not particularly well suited for data recording or automatic control applications. With some degree of modification, it is possible to use the liquid expansion principle to produce mechanical or physical changes. Filled temperature-measuring instruments such as the one shown in Figure 10-2 are the result of this modification.

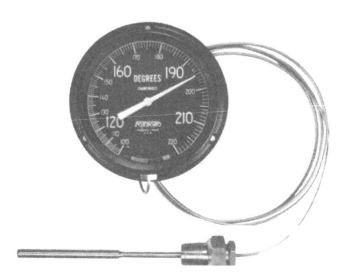

Figure 10-2. Filled-tube thermometer. (*Courtesy of The Foxboro Company.***)**

Filled temperature-measuring instruments consist of a sensitive pressure bulb and a thin metal tube or capillary connected to a bourdon element. Liquid or gas sealed inside the system will expand with an increase in temperature. Fluid expansion, in turn, causes an increase in inside system pressure. Corresponding changes in pressure are transmitted to the Bourdon element through the capillary tube. The Bourdon element is essentially a coil of flat tubing with its extreme end sealed. Increases in pressure have a tendency to cause the element to unwind or become straight. This physical change can be used to indicate temperature values on a chart or calibrated scale.

Filled temperature recorders usually employ a circular chart and a calibrated scale with a recording pen attached to the pressure element. A partial cutaway view of a filled temperature recorder is shown in Figure 10-3. The spiral coil near the center of the instrument is the Bourdon element, which is attached to the chart recording mechanism. The sensitive pressure bulb may be located some dis-

Figure 10-3. Partial cutaway view of a filled temperature recorder. (*Courtesy of The Foxboro Company.*)

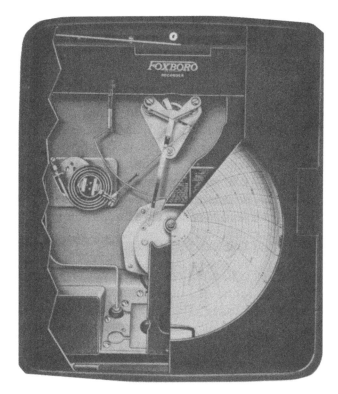

tance away from the recorder unit. Connection of the sensing bulb to the Bourdon element is achieved by the capillary tube. Filled system recorders can indicate temperature values for 24 hours or for a 7-day operational period. As a general rule, filled system equipment requires periodic maintenance when in service.

Bimetallic Thermometers

Bimetallic thermometers are nonelectric instruments that make use of the principle of the coefficient of linear expansion. Fabrication of this instrument is achieved by bonding together two pieces of metal that have different linear expansion coefficients. An increase in temperature will cause greater expansion of the metal strip with the highest expansion coefficient. Bonding of the low expansion strip to the high expansion strip will cause physical distortion to occur with changes in temperature. The composite strip will bend in the direction of the metal with the lower expansion rate. The resulting amount of deflection that occurs is proportional to square of the length, the total change in temperature, and is inversely related to metal thickness.

In order to generate a rather suitable physical change by the expansion principle, the bimetal element should be as long as possible. In practice, the element is generally formed into a flat spiral or single helix coil. The outermost part of the element is attached to the housing with the loose end attached to an indicating hand. An increase in applied temperature causes the coil element to wind up in the clockwise direction. Movement of the indicating hand registers temperature values on a calibrated scale.

Figure 10-4 shows the spiral element of a bimetallic thermometer. The sensing element of this instrument is exposed for quick response to surface temperature measurements. Permanent magnets attached to the housing are used to secure the thermometer to a metal surface. In some applications the unit may be permanently attached to a surface for continuous temperature monitoring. Temperature scales and ranges are available in a wide variety of values and sizes. This type of instrument is fairly accurate, rather inexpensive, and requires very little maintenance for long periods of operation.

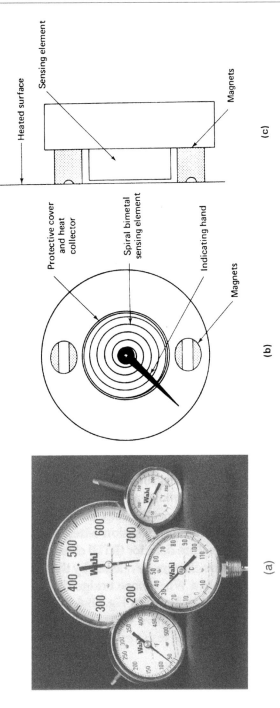

Figure 10-4. (a) Bimetallic thermometer; (b) back view; (c) side view; *(Courtesy of Wahl Instruments Inc.)*

ELECTRONIC TEMPERATURE INSTRUMENTS

Electronic instruments at one time were rarely used in the evaluation and control of building systems. The expense of the instrument and precision of its measuring capabilities were not particularly needed or suited for most equipment applications. However, technological developments have caused a decided change in electronic instrumentation for building system applications.

Electronic temperature measuring instruments are unique when compared with nonelectrical measuring equipment. First, electrical energy must be supplied to the instrument by an external source in order for it to be operational. Second, it must employ a special sensing element, known as a *transducer,* to change temperature variations into electrical signals. Transducers are usually small in mass and can be located some distance away from the instrument. In addition to this, electronic instruments can be easily made portable, which increases their versatility and measuring techniques.

Electronic instruments are generally classified according to the operational principle of the temperature-sensing element. Divisions include resistance changes, voltage generation, radiation detection, and optical comparisons. The transducer of the instrument must have the ability to distinguish between temperature changes in a given space or area and be able to measure the amount of heat in a specific object or body.

Resistance Temperature Instruments

Resistance-temperature-instrument operation is based on the property of certain metals to change resistance when subjected to heat. The sensing element of this instrument in simply a long piece of wire formed into a coil and wound around a ceramic core. The entire assembly is then enclosed in a protective metal sheath. Electrical connection is made to the wire coil by terminals or extension wires. A cross-sectional view of a resistance temperature detector (RTD) element is shown in Figure 10-5.

Temperature changes developed by the RTD element are obtained by connecting it to a Wheatstone bridge circuit. A representative two-wire bridge circuit is shown in Figure 10-6. Note that there is a source of electrical energy, a sensitive current-measuring instrument called a

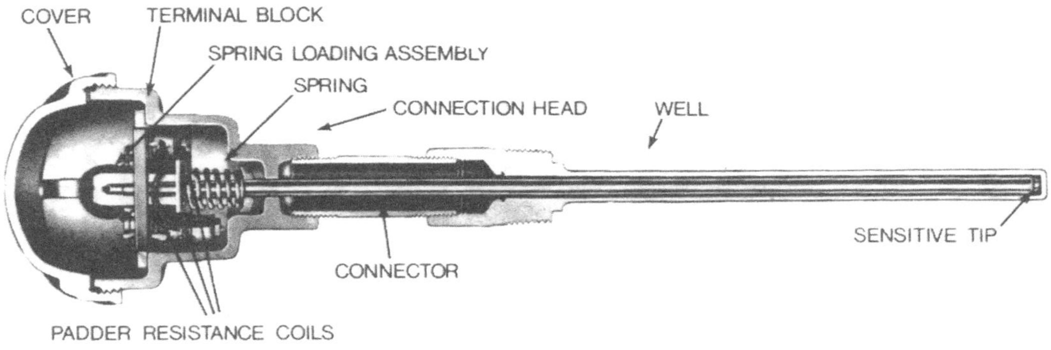

Figure 10-5. Cross-sectional view of a resistance temperature detector (*Courtesy of The Foxboro Company.*)

SENSITIVE TIP

WELL

CONNECTION HEAD

SPRING LOADING ASSEMBLY

SPRING

CONNECTOR

TERMINAL BLOCK

PADDER RESISTANCE COILS

COVER

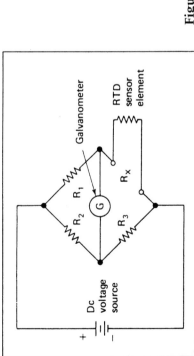

Figure 10-6. Two-wire Wheatstone bridge.

Galvanometer

R_1

R_2

R_3

R_x

RTD sensor element

Dc voltage source

galvanometer (G), three resistors, and the RTD element.

When electrical energy is applied to a bridge circuit, current flows from the source and divides into the two resistance legs. Then the resistance of the R_2-R_3 leg equals that of the R_1-RTD leg, an equal amount of current will flow in each leg. This action will produce an electrical balance which is indicated by a zero reading on the galvanometer.

In order to determine a specific temperature by the bridge circuit, the resistance of the RTD must be determined. In practice, R3 is normally made variable so that it can be adjusted to match the resistance of the RTD element. At balance, the resistance of RTD is determined by the formula

$$\text{RTD} = \frac{R_1}{R_2} \times R_3$$

When the resistance value of the RTD is decided, temperature can be determined by locating this value on a resistance-temperature graph. Platinum wire, for example, offers 10 ohms of resistance at 0°C. An increase in temperature to 100°C will cause a linear rise in resistance to 13.8 ohms. Temperature can therefore be determined by measured changes in the resistance of the RTD element.

A number of portable bridge temperature-measuring instruments are available today. Temperature is measured by simply adjusting a calibrated resistor to produce a balanced condition of the bridge. Resistance registered on the calibrated dial is indicative of the sensed RTD value. This value located on an appropriate resistance/temperature scale will indicate the temperature sensed by the RTD element.

A rather recent development to the RTD family of temperature-measuring instruments is the portable digital thermometer. Figure 10-7 shows a representative digital thermometer that will measure temperatures over the range –50 to 900°F (–45.5 to 482.2°C). When the instrument is equipped with a platinum RTD probe its accuracy is 0.1°F or 0.05°C.

Operation of the digital thermometer is achieved by first energizing the electrical components by the on/off switch. Placing RTD sensor probe on a surface or in a prescribed area will produce a stabilized reading in a few seconds. Temperature readings are indicated directly on the instruments digital display in values of 0.1 degree. In some instruments, the display will indicate either Celsius or Fahrenheit

Figure 10-7. Portable digital thermometer with RTD probe. (*Courtesy of Wahl Instruments Inc.*)

temperature according to the °F or °C select position. The versatility of this instrument, plus its operational simplicity and accuracy, make it a very popular temperature-measuring instrument.

Thermocouple Instruments

Thermocouple instruments respond to a rather simple and versatile principle based on the generation of electricity by heat. When two dissimilar metals wires are joined together at a common point and subjected to heat, a DC voltage is produced at the free ends of the wire. The resulting voltage measured across the two wires can be used to indicate the temperature applied to the measuring junction. Figure 10-8 shows a simplified example of a thermocouple measuring instrument. In practice, this type of instrument is usually called a millivolt pyrometer or simply a pyrometer.

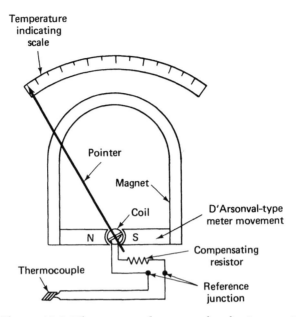

Figure 10-8. Thermocouple measuring instrument.

The amount of voltage developed by a thermocouple is generally quite small. Representative values are in the range 0 to 70 millivolts for temperature values from 0 to 3000°F (–19 to 1634. 7°C). A millivolt is 1 one-thousandth of a volt (0.001 V).

The physical components of a millivolt pyrometer are a d'Arsonval type of electric meter movement, a compensating resistor, and the thermocouple. Heat applied to the thermocouple measuring-junction will cause a corresponding voltage to appear at the reference junction. A resulting current will flow through the meter coil, producing a corresponding electromagnetic field. An interaction between the coil field and the permanent magnet field of the meter movement will cause a proportionate deflection of the movable coil. An indicating hand attached to the moving coil will display the resulting amount of deflection in millivolts. A scale attached to the meter is used to indicate temperature in °C or °F.

Digital thermometers that employ thermocouples as sensors are finding widespread acceptance in the building maintenance field. This type of instrument is easy to use, produces accurate indications of temperature, has a rapid response time, and is quite rugged in construction. A simplification of the electronics of this instrument is shown by the block diagram of Figure 10-9. Voltage developed by the thermocouple varies in value according to temperature applied to its sensor probe. Measurable quantities of this type are normally described as analog values. Analog voltage is then changed into digital information by the analog-to-digital converter. The corresponding digital signal is counted and applied to the digital decoder driver circuit. The signal is then converted into numbers that are displayed on the readout as indications of temperature. A great deal of the electronic circuitry of this instrument is achieved by microminiature components called *integrated circuits*.

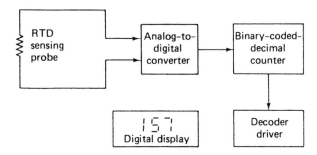

Figure 10-9. Electronic parts of a digital instrument.

The digital thermometer of Figure 10-10 is capable of measuring a rather wide range of temperatures according to the thermocouple selected. Typical values are from –99.8° to +999.8°C or °F with a 0.2°F or °C display resolution. The full range of temperature is achieved by four different thermocouples that cover various parts of the range. Instruments of this type are powered by either AC line voltage or a rechargeable battery pack, according to its application.

Thermistor Instruments

Thermistors are very small solid-state devices that are used in instruments to detect changes in temperature. The solid-state designation associated with a thermistor implies that it is made of a semiconductive material. Fabrication is normally achieved by mixing together such things as manganese, cobalt, and nickel oxides. Thermistors are made in the shape of beads, rods, washers, and various other configurations.

Figure 10-10. Portable thermocouple digital thermometer. (*Courtesy of Omega Engineering Inc.*)

For temperature instruments, the thermistor is placed into a variety of probe structures. Figure 10-11 shows some of the thermistor shapes available for use in instrument probes.

A distinguishing feature of the thermistor is its classification as a negative-temperature-coefficient device. This characteristic refers to its ability to change resistance in a direction that is opposite to that of the temperature being sensed. An increase in temperature causes a decrease in thermistor resistance, whereas a decrease in temperature causes a increase in resistance. This characteristic is opposite to that of an RTD, which has a positive temperature coefficient.

Thermistor instruments measure temperature by monitoring electrical circuit changes in current and voltage. The circuit of Figure 10-12 shows a simplification of a representative thermistor temperature instrument. Notice that the circuit employs a DC energy source, a variable resistor, a thermistor, and a very sensitive microammeter. A microammeter will measure 1/1,000,000 or 0.000001 of an ampere of current.

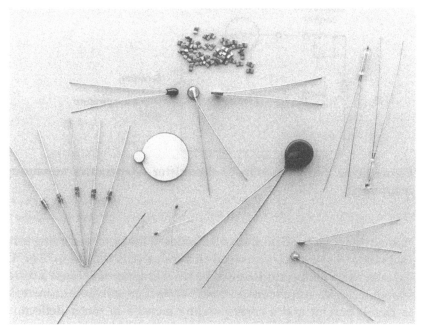

Figure 10-11. Thermistor probes. (*Courtesy of Fenwal Inc.*)

Operation of the thermistor circuit involves a calibration function and a measurement function. In most instruments the calibration function is achieved in the factory when the instrument is being assembled. Adjustment of the calibration resistor produces a representative current flow in the circuit that is indicative of the resistance of the thermistor at a specific temperature value.

When the thermistor circuit is energized by the DC source, it is ready to perform the measurement function. Current now from the source through the circuit will produce a deflection of the meter according to its calibration. Meter detection is read on a scale that is marked in °F or °C. An increase in temperature sensed by the thermistor will cause a corresponding decrease in circuit resistance. This action will cause an increase in circuit current and a corresponding increase in meter deflection. Deflection of the meter will show a representative increase in temperature on the indicating scale. A decrease in temperature sensed by the thermistor will cause an increase in circuit resistance. This action will cause a decrease in current flow and reduced deflection of the meter.

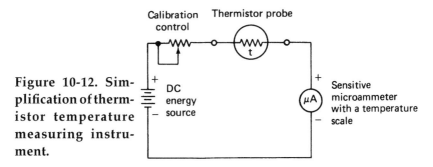

Figure 10-12. Simplification of thermistor temperature measuring instrument.

Thermistor-temperature-measuring instruments are very sensitive, stable, fast responding, and employ rather simple circuitry. The length of probe leads does not adversely affect circuit operation, and there is no specific polarity to the probe. The uniqueness of a thermistor instrument to produce changes in resistance that are almost entirely a function of temperature makes it a vital measuring device in the evaluation of building systems.

Radiation Pyrometers

A number of everyday experiences tells us that hot objects have a tendency to send out or radiate heat waves. We can feel this radiation on our skin or face without touching the warm object. Infrared radiation, which is the technical way of describing this type of energy, travels through the air by invisible electromagnetic waves. The behavior of this energy is very similar to light. It travels at the speed of light but has a significantly shorter wavelength.

Instruments that respond to the amount of thermal energy radiated from the surface of an object are called radiation pyrometers. Temperature measurements made by this type of instrument are achieved without making direct contact with the object being tested. A special optical component is employed that will permit infrared energy to be focused on its active surface by a lens. The detector of this component changes heat energy into an electrical signal. Amplification of the signal is then needed before it can be applied effectively to the measuring component.

A version of the radiation pyrometer is shown in Figure 10-13. Operation is based upon analog signals being converted into digital information, which is then counted and decoded before being applied to

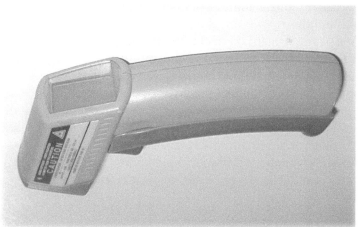

Figure 10-13. Digital infrared thermometers *(courtesy of Mikron Instrument Co.;*
b, courtesy of Wahl Instruments Inc.)

the readout display. The measurement range of this instrument is 0° to 100°C or °F in 1-degree scale divisions. Measurements can be obtained over a working distance of approximately 20 feet or more. Instruments of this type are ideally suited for measurement of comfort heating system operation and heat losses.

HUMIDITY MEASUREMENT

Control of the moisture content of air is a very important consideration in the evaluation of air-conditioning equipment and heating-system effectiveness. It has been a recognized fact for years that relative humidity (RH) is an essential part of a healthful and comfortable environment. In addition to this, inside building humidity also has a great deal to do with the operational efficiency of building environment equipment. If the temperature of a room or space is increased without adding moisture to the air, the relative humidity will normally decrease. Dry RH levels have a tendency to cause a person to feel somewhat uncomfortable. To feel more comfortable, it is usually necessary to raise the ambient temperature of the room to a higher level. This obviously calls for longer equipment operational cycles and increased energy consumption.

The moisture content of air during the summer has a tendency to increase rather substantially with temperature in many parts of the country. Hot, humid conditions normally cause a person to feel very uncomfortable. Perspiration will not evaporate effectively into the air, which causes a person to have a muggy or sticky feeling. Lowering the inside temperature of a building has a natural tendency to reduce the moisture content of air. Reduced RH levels normally cause increased comfort and less operational time for air-conditioning equipment. System efficiency is influenced a great deal by the RH level of a building.

Sling Psychrometer

The relative humidity of air is commonly measured with a sling psychrometer. This instrument consists of two identical thermometers mounted on a light frame that is designed to be whirled in the air. One thermometer measures the dry-bulb (DB) temperature of the air. The second thermometer has its mercury bulb surrounded by a cloth wick that is saturated with water. Wet-bulb (WB) temperature is read on this thermometer. When the entire assembly is whirled or "slung" through the air, evaporation of the wet wick cools the WB thermometer. A very dry atmosphere or one with a low relative humidity will cause the wick to evaporate very rapidly. This causes a very large difference between WB and DB temperature values. A large difference in temperature indicates a rather low value of relative humidity. Higher RH values are indicated by a very small difference in WB and DB values. At 100% RH the wet-bulb and dry-bulb temperatures are equal. This indicates when the air is fully saturated with moisture and that it will not accept any more. The wick moisture of the WB thermometer will not evaporate effectively at saturation, thus causing it to have the same temperature value as the DB thermometer.

A compact sling psychrometer is shown in Figure 10-14(a). This particular assembly measures wet- and dry-bulb temperatures of the surrounding air and permits immediate conversion of the readings to relative humidity. After whirling the sling in the air for approximately 1 minute, it is returned to the housing as shown in Figure 10-14(b). The dry-bulb temperature is aligned with an indicating arrow on the housing. Wet-bulb temperature values on the sling indicate the value of relative humidity on the etched scale. RH values of this type are accurate to within 1% of the scale value. For greater precision WB and DB temperature values can be plotted on a relative humidity table similar to the one shown in Figure 10-15.

Mechanical Humidity Instruments

Mechanical humidity instruments employ any one of several materials that change their physical dimensions when moisture is absorbed. Human hair, horsehair, and animal membranes are frequently used to accomplish this operation. Human hair, for example, changes its length by 2-1/2% from 0 to 100% RH. This physical change can be mechanically amplified by gears and levers to move an indicating l-and over a graduated scale. A direct-reading mechanical humidity instrument is shown in Figure 10-16.

Operation of a mechanical humidity indicator is based upon the circulation of moist air through an enclosed chamber. Most instruments of this type rely upon gravity feed for circulation of the air being evaluated. Response time of the instrument is somewhat slow compared

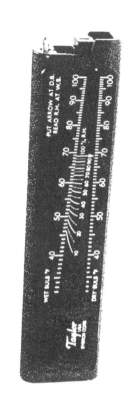

(b)

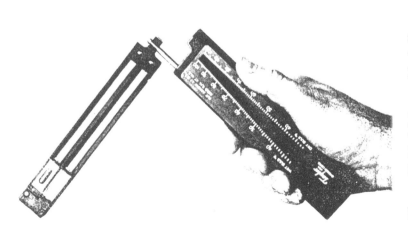

(a)

Figure 10-14. (a) Assembled sling psychrometer; (b) psychrometer placed in housing for reading. (*Courtesy of ABB Kent-Taylor.*)

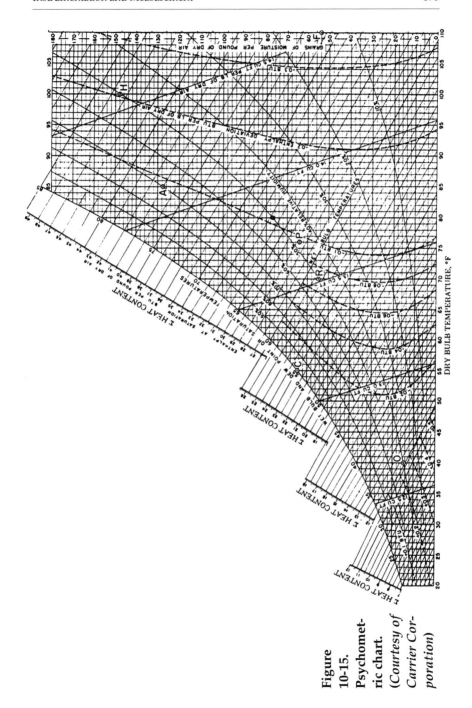

Figure 10-15. Psychometric chart. (Courtesy of Carrier Corporation)

**Figure 10-16. Direct-reading humid-
ity instrument/thermometer. (***Courtesy of AABBEON Cal, Inc.***)**

with other RH measuring techniques. This type of instrument is fairly reliable for measurements between 15 and 90% RH. The low cost of the instrument, its operational simplicity, and the advantage of direct-reading indications cause it to be widely used in building equipment evaluation applications.

Electronic Hygrometers

A number of very specialized moisture sensors are used with electronic hygrometers to measure the moisture content of air. These sensors are designed to react to different levels of moisture by causing a physical change in some electrical property. For the most part, sensors of this type are rugged, compact, and can provide remote readout capabilities that cannot be achieved by other RH instruments. A representative RH sensor probe is shown in Figure 10-17.

RH probes are made of an insulating material that is interleaved with a grid-like structure. The grid is coated with a hygroscopic material such as lithium chloride. Hygroscopic materials have the inherent ability to absorb and release moisture. Moisture absorbed by the sensor causes a decided change in its electrical resistance. This electrical change, when used in one leg of a Wheatstone bridge, will produce a significant change in circuit current. Variations in bridge current are then used to indicate relative humidity values on a calibrated meter scale. Probes of this type also employ a resistance thermometer that will respond to a given range of temperature. Temperature and humidity are both indicated by the readout part of the instrument.

The electronic circuitry of an RH meter is usually housed in the readout or display section of the instrument. Resistance changes in the probe are applied to a very sensitive alternating-current bridge circuit.

Figure 10-17. Relative-humidity probes. (*Courtesy of General Eastern Instruments.*)

The resulting AC output signal is amplified and linearized before being applied to the readout or display part of the instrument. Display can either be of the hand-deflected meter type or have a digital readout, as shown in Figure 10-18. These instruments will display RH for a range of 0 to 100% or temperature from –50 to + 150°F.

PRESSURE MEASUREMENT

The operation of some building equipment is based upon pressure changes that occur at different parts of the system. Measurements of this type are used to evaluate equipment performance, operational efficiency testing, and in routine maintenance procedures. Some pressure measurements are performed on a continuous basis, whereas others are made temporarily.

Air-conditioning equipment and airflow are a few of the common applications of pressure measurement. A person working with equipment should be familiar with pressure-measuring instruments.

Pressure Gauges

A large number of the pressure gauges in operation today respond to the elastic deformation principle. These instruments employ

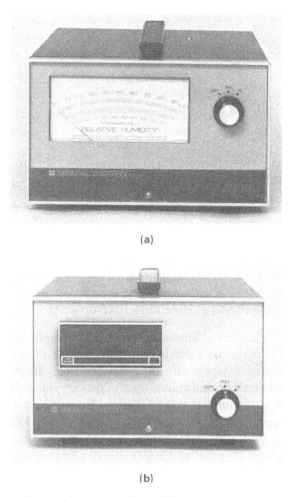

(a)

(b)

Figure 10-18. Electronic relative-humidity instruments: (a) meter type; (b) digital type. (*Courtesy of General Eastern Instruments.*)

an element that physically changes shape when different pressure values are applied. Figure 10-19 shows three very common elements that are used to produce pressure indications. The spiral and helix coil have a natural tendency to uncoil when pressure is applied. The Bourdon tube responds to pressure by becoming straight. The physical reaction to pressure produced by any of these elements can be harnessed to move an indicator hand or drive a stylus on a paper chart.

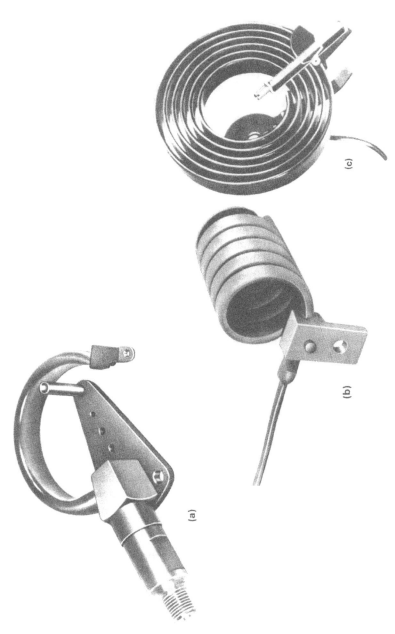

Figure 10-19. Types of pressure elements: (a) Bourdon; (b) spiral; (c) helix. (*Courtesy of Honeywell, Inc.*)

A simplification of the components of a pressure gauge employing the Bourdon tube element is shown in Figure 10-20. Pressure applied to this instrument causes the Bourdon element to straighten. This physical change in linked to a sector gear and pinion. An indicating hand attached to the pinion gear causes it to move physically with changes in pressure. An increase in pressure causes clockwise rotation of the pinion and indicating pointer. Reduced pressure causes the element to return to its original shape. This action causes counterclockwise rotation of the pinion gear and indicating pointer. An assembled Bourdon pressure gauge is shown in Figure 10-21.

Manometers

A variety of manometers are available today for different pressure-measuring applications. The operational principle of this type of instrument is basic to all manometers. The U-tube manometer of Figure 10-22 is very popular and illustrates the simplicity of this operating principle.

In its simplest form a hollow glass or plastic tube is formed into a U shape with the open ends of the tube pointing upward. The tube is then half filled with a fluid such as water or mercury. With both ends of the tube open, the liquid will balance itself at the same height in each leg. The represents a zero indication of pressure. A scale placed in the center of the U-tube could be positioned to indicate the zero reference point.

When a positive pressure is directed into one leg of the U-tube it forces the liquid down in that leg and up in the other leg. The resulting difference in the height of the two liquid columns indicates the pressure. An indication of pressure would be in inches of water or inches of mercury. This could be read directly on a scale positioned in the center of the U-tube. Manometer scales are based upon the pressure required to move the rill liquid. One inch of mercury (Hg) is equal to 0.492 pounds per square inch (psi). Water, being much lighter in weight than mercury, produces a greater change in column height. One inch of water is equivalent to 0.036 psi of pressure or 1 psi = 2.31 ft of water. A convenient flexible-tube manometer and a portable roll-up type of manometer are shown in Figure 10-23.

Indicator pointer
Bourdon element
Sector gear
Link
Pinion gear
Socket
Filled tube

Figure 10-20. Components of Bourdon-rube pressure gauge. (*Courtesy of Marshalltown Instruments.*)

0
10
20
30
40
50
60
70
80
90
100
LBS. PER SQ. IN.
MODEL PA
FOXBORO
FOXBORO, MASS.
U. S. A.

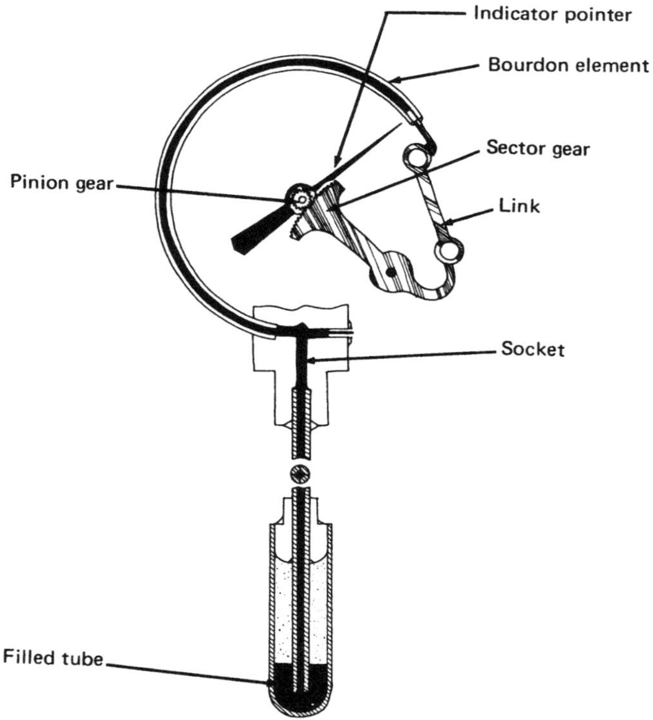

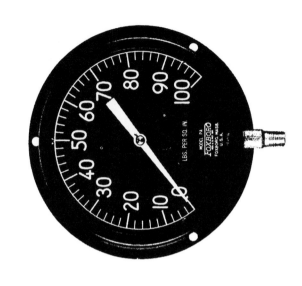

Figure 10-21. Assembled pressure gauge. (*Courtesy of The Foxboro Company.*)

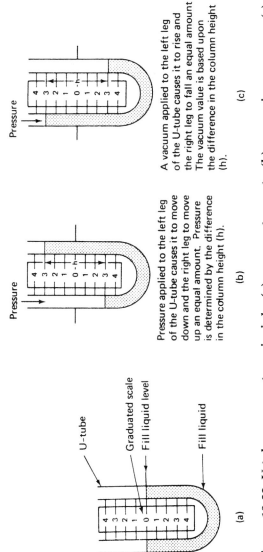

Pressure

Pressure applied to the left leg
of the U-tube causes it to move
down and the right leg to move
up an equal amount. Pressure
is determined by the difference
in the column height (h).

(b)

Pressure

A vacuum applied to the left leg
of the U-tube causes it to rise and
the right leg to fall an equal amount.
The vacuum value is based upon
the difference in the column height
(h).

(c)

U-tube

Graduated scale
Fill liquid level

Fill liquid

(a)

Figure 10-22. U-tube manometer principle: (a) manometer parts; (b) measuring pressure; (c) measuring vacuum.

ELECTRICAL MEASUREMENT AND INSTRUMENTATION

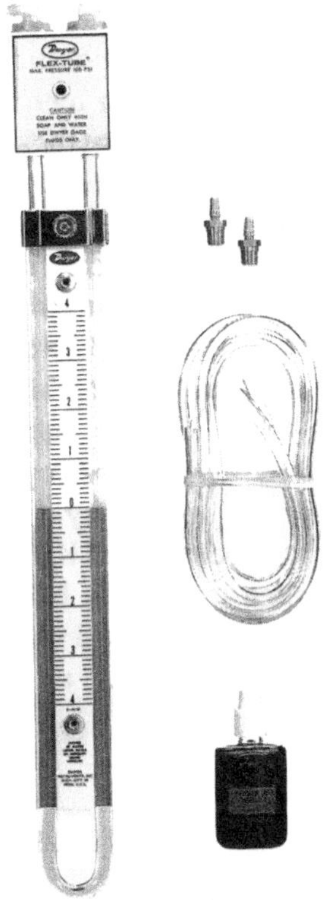

Electricity is involved in the operation of nearly all building equipment today. In some applications, such as electric heating and lighting, it serves as the primary source of energy for system operation. In other applications it provides the energy needed to control specific system functions. The electrically energized thermostat of a gas-burning furnace or air-conditioning system is an example of the control function. Electricity is also supplied to motors that are used to control airflow, circulate hot water, or actuate the compressor of an air conditioner or heat pump. Electrical measurement plays a very important role in the operation and evaluation of building equipment.

Electrical instruments that are used to perform the measurement function are either energized by the system being tested or by an independent source of electricity. Voltmeters, ammeters, wattmeters, and kilowatt hour meters derive their operational energy from the system under test. Specialized instruments that employ active components such as transistors vacuum tubes or digital displays must also be energized by an independent electrical source in order to function.

Figure 10-23. (a) Flexible-tube manometer; (b) portable roll-up manometer. (*Courtesy of Dwyer Instruments Inc.*)

Electrical instruments are designed primarily to measure specific electrical values and display this information in a usable manner. Hand-defection meters digital display instruments, and chart recording instruments may be used to perform the measurement function. These instruments may be designed to measure only one specific electrical quantity over a rather narrow range or several different quantities with multiple ranges. Typical electrical quantities that are measured in the evaluation of building equipment are voltage, current, power, resistance, and power factor.

The measurement of certain electrical quantities often causes some confusion because of the terminology used. Current, for example, is measured with an ammeter. The fundamental unit of current is the *ampere* (A). Voltmeters are used to measure electrical voltage. The fundamental measuring unit of voltage is the *volt* (V). Electrical power is somewhat unusual because it is measured in *watts* (W). *Watthours* (Wh) are a measure of the power consumed for a given unit of time. Resistance is measured with an ohmmeter. Meters of this type may have the word "ohms" or the Greek letter omega (Ω) displayed on the scale. Power factor is measured with an instrument that has a scale graduated in values from 0 to 1. A power factor (PF) of 1 is ideal, with decimal values less than 1 being less desirable.

Hand-Deflection Instruments

Electrical instruments that rely upon the movement of an indicating hand or pointer are referred to as hand-deflection instruments. The volt-ohm-milliammeter (VOM) of Figure 10-24 is a typical example of a hand-deflection instrument. This particular instrument is a multifunction, multirange meter. Single-function instruments that employ only one measurement range are also available using the same principle of operation.

The fundamental operation of a hand-deflection instrument is based upon the mechanical action of its meter movement. A simplification of the basic permanent magnet or d'Arsonval meter movement construction is shown in Figure 10-25. The moving coil assembly of this instrument is

Figure 10-24. Hand-deflection VOM. (*Courtesy of The Vector Group Inc., Instrument Div.*)

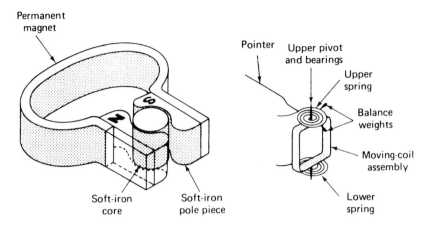

Figure 10-25. Components of a d'Arsonval meter movement construction.

placed between two soft-iron pole pieces that are attached to the permanent magnet. Physical deflection of the moving coil is based on an interaction between the permanent magnet field and the electromagnetic field. The moving coil is physically oriented so that the indicating hand comes to rest on the left side of the movement when it is not energized.

When the test probes of a meter are connected to an energized electrical circuit, a small amount of current will flow from the circuit through the coil assembly. A resulting electromagnetic field will immediately appear around the coil. Interaction between the permanent magnetic field and the electromagnetic field will cause the coil to move proportionately according to its field strength. This reaction will cause the indicating hand to deflect to the right of its normal resting position. Figure 10-26 shows a simplification of meter movement operation.

In a sense, the basic hand-deflection meter movement is considered to be a very sensitive current-measuring instrument. Deflection of the moving coil is based on an extremely small amount of current that is applied to it. The resistance of the coil is very low, with typical values being approximately 100 ohms. The basic meter movement may be modified to measure large amounts of current, voltage, or resistance.

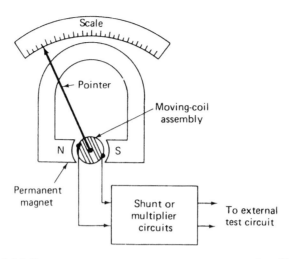

Figure 10-26. Permanent-magnet meter movement simplification.

Digital Instruments

Nearly all electrical quantities can be measured by a digital instrument. The particular quantity being measured is limited only by the capabilities of the input device. Voltage, current, resistance, watts, light level, power factor, and temperature are some of the common applications of digital instruments.

With such a wide range of digital instrument applications. one would think that this type of equipment would be quite complex. On the contrary—nearly all digital instruments employ a number of common elements. The digital display or readout unit is primarily the same for all instruments. In addition to this, the counter assembly is nearly always of the binary-coded-decimal (BCD) type. Its output is used to drive a decoder of the seven-segment type. The counter, decoder-driver, and readout are primarily responsible for the counting function of the digital instrument.

Figure 10-27 shows the essential parts of the counter of a digital instrument. In operation, a two-state (on or off) or binary signal is applied to the input of the counter assembly. Each pulse or "on" state of the signal causes an advance of one count. The output of the counter is fed into the BED decoder, which drives the display device. Zero to nine pulses would be received by the first BED counter and cause a corresponding number 0 to 9 to be displayed on the 1s output of the

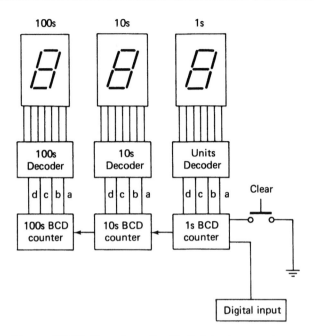

Figure 10-27. Counter of a digital instrument.

display. A count of 10 or more pulses would activate both the 1s and 10s counters. After the ninth count a 1 would appear on the 10s counter and a zero on the 1s counter. Values up to 99 could be displayed on the two counters. Something in excess of a count of 99 will appear on the 100s, 10s, and 1s display. This part of the instrument is quite simple, very accurate, and relatively inexpensive to accomplish. The display part of nearly all digital instruments operates on this basic principle.

The input section of a digital instrument is undoubtedly the most unique part of the entire system. Special analog-to-digital (A/D) converters are used to achieve this function. This section of a digital instrument is designed to change analog information such as voltage or current values into binary or two state information. Figure 10-28 shows a block diagram of a digital instrument with the input function attached to the counter section.

Digital Volt-Ohm-Milliampere Meter. One of the most widely used digital instruments today is the digital volt-ohm-milliammeter (DVOM). This instrument is designed primarily to change analog volt-

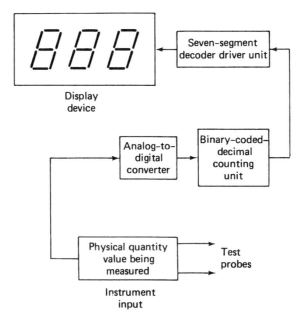

Figure 10-28. Schematic of a digital instrument.

age, current, or resistance values into binary signals that can be counted and displayed in digital form. Measurement is achieved by A-to-D conversion then by counting and display. Figure 10-29 shows two digital VOMs that operate through this process. Modern meters are available in smaller sizes with auto-ranging capability.

The A-to-D converter of a digital instrument represents the most unusual part of the entire instrument. This particular function is responsible for converting unknown analog values into usable digital signals. When performing this function, the converter must be accurate, capable of rejecting extraneous noise, and being fairly stable over a suitable operating range. A number of different A-to-D conversion techniques are capable of achieving this operation today. Included in this are such things as voltage-controlled oscillators, ramp function conversion, and dual slope conversion. Only one of these conversion techniques, the voltage-controlled oscillator method, will be presented in this discussion.

The voltage-controlled oscillator (VCO) method of A-to-D conversion is one of the simplest of all conversion techniques. It relies on

Figure 10-29. Digital volt-ohm-milliampmeters. (*Courtesy of Fluke Corporation.*)

the ability of a special oscillator to change frequency when an input voltage is applied. The output frequency of the oscillator is then counted for a fixed interval of time and displayed on the digital readout.

Figure 10-30 shows a simplified block diagram of a VCO instrument. Operation is based on the frequency of the VCO being scaled so that it will produce a certain output for each volt of input. One volt, ohm, or ampere of input applied to the VCO should produce a 1000 count on the display. For a digital VOM with a power-line time-base generator of 60 Hz, the VCO would need to operate in a range 0 to 60,000 Hz. This would be described as a 60,000 Hz/V sensitivity.

The time-base generator of a digital VOM is responsible for turning on or "gating" the output of the VCO for a specific time. In practice, many digital instruments employ the 60-Hz power-line frequency to trigger the time-base generator into operation. With a 60-Hz sine wave applied to its input, the output will be a 60-Hz square wave, as shown in Figure 10-30. The output of the time-base generator

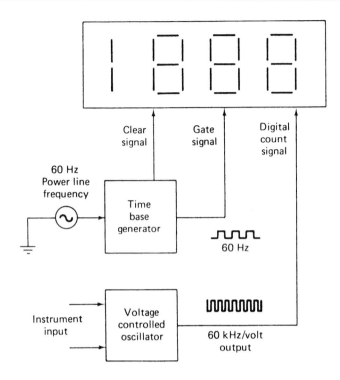

Figure 10-30. Schematic of a voltage-controlled oscillator DVOM.

is usually called the gate signal.

Assume now that 1 V is to be measured by our basic VCO digital VOM. When the input to the VCO is 1 V, it causes the frequency of the 60,000-Hz/V oscillator to advance to 60,000 Hz. The time-base generator connected to the 60-Hz power line will also receive a square-wave gate signal of 60 Hz at the same time. This action will cause the gate to turn on every 1/60 second. During an on interval, 1/60 of 60,000 would produce a count of 1000. A decimal-point control circuit would be energized, causing the display to indicate a decimal value of 1.000 or 1 V. After a short period of time the counter would reset to zero and start a new count. If a different voltage appears at the input, it would make the necessary VCO frequency changes to generate a new count. No change in the input would cause a new 1-V indication to appear on the readout. In a strict sense, the DVOM simply counts the VCO output and displays it on the readout.

Clamp-Type Instruments

Clamp-type instruments are designed to measure the amount of AC current flowing in a power line without disturbing circuit operation. The clamp-type instrument of Figure 10-31 has movable tongs that can be opened and placed around a current-carrying electrical conductor. Changes in current through the conductor cause it to develop an electromagnetic field. When this field cuts across the closed clamp, it causes a corresponding current to be induced into the clamp coil. A simplified circuit of a clamp-type ammeter is shown in Figure 10-32.

When using the clamp-type instrument to measure a current value, only one conductor of the circuit should be placed inside the clamp. This type of ammeter is designed to respond only to alternating-current electricity. Ordinarily, clamp-type instruments are also equipped to measure voltage and resistance values. These additional functions utilize external test leads so that the instrument can be used to make measurements without the clamp.

Figure 10-31. Clamp-type instrument. (*Courtesy of Fluke Corporation.*)

Power-Measuring Instruments

Electrical power is measured with a rather unusual instrument called a wattmeter. This particular instrument employs a special type of meter movement that is called a dynamometer. Figure 10-33 shows a simplification of the dynamometer meter movement. Note that the basic meter has electromagnetic field coils instead of the conventional permanent magnetic field. The moving coil of this movement is similar in many respects to the basic d'Arsonval type of meter of the VOM. Both sets of coils of the dynamometer movement derive operational

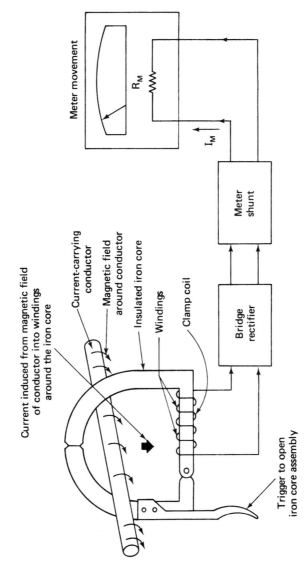

Figure 10-32. Schematic of a clamp-type ammeter.

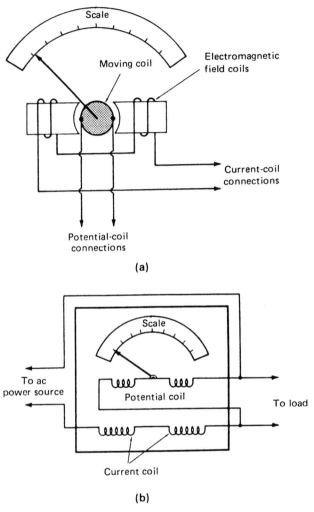

Figure 10-33. Basic dynamometer: (a) meter movement; (b) circuit.

energy from the circuit under test.

The two sets of coils of a dynamometer perform different functions in the operation of the instrument. The current coils produce a stationary magnetic field when they are energized. This coil set is connected in series with the circuit under test and performs the current function of the instrument. The movable coil is high resistant and serves as the voltage-measuring function. It also derives its operational energy from circuit under test. The independent voltage coil is connected in parallel or across the line voltage of the circuit.

When the current coil and voltage coil of a dynamometer are both connected to an energized circuit, each will produce a corresponding electromagnetic field. The resulting movement of the indicating hand is the product of the two independent fields. With one field resulting from circuit current and the other from voltage, deflection indicates the product of current and voltage or watts. When a wattmeter is connected to a circuit, it is imperative that both the current and voltage coils be properly attached. The current coil must be connected in series by breaking the circuit and inserting the meter between the two connection points. The voltmeter coil must be connected across the source voltage or in parallel with the line voltage. No deflection will occur if only one coil is connected independently.

A rather recent addition to the power meter field is the power analyzer of Figure 10-34 This instrument will independently measure current, voltage, and watts and indicate the values on a digital display. Independent circuit connections must be made for both current and voltage values. The watts indication of this instrument is achieved electronically by sampling the voltage and current values. The instrument first converts analog current and voltage values into digital information then counts these data for display on the readout. The A-to-D converter and counting functions are similar to those used in the DVOM.

Power-Factor Measurement

Power factor (PF) is described as ratio of true power to apparent power. True power, which is measured in watts (W), is an indication of the actual power being converted into work by a particular component or system. Apparent power, which is measured in volt-amperes (VA), is an indication of the power being delivered to a system. A power factor of 1.0, or 100%, is ideal for good operational efficiency. As

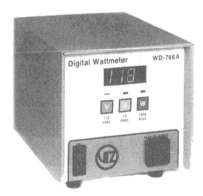

Figure 10-34. Power analyzer. (*Courtesy of The Vector Group Inc., Instruments Div.*)

a rule, only pure resistive electrical loads are capable of developing a unity power factor. It is important to monitor the power factor of an electrical system in order to examine its efficiency. This has a great deal to do with the amount of power that is being turned into work and that which is actually being supplied by the system.

The operational principle of a power-factor meter is shown in Figure 10-35. This particular instrument is very similar in construction to the hand-deflecting wattmeter discussed previously. An apparent difference in the power-factor meter is the construction of its movable coil. As noted, two coils are mounted on the same movable shaft. Rotation is dependent on the resulting field strength of the two coils. Physically, the coils are mounted so that they are displaced 90°. One coil connected to the AC power line is in series with a resistor; the second coil is connected in series with an inductor. The resistor-coil connection represents the in-phase component of the AC power line, and the in-

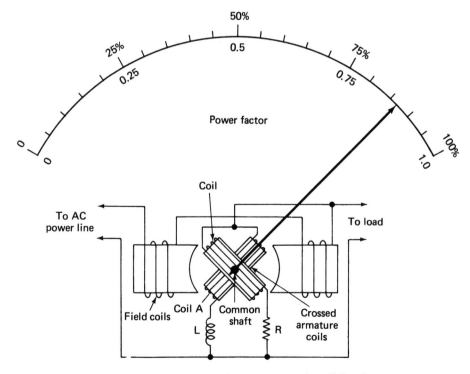

Figure 10-35. Power-factor-meter simplification.

ductor-coil connection represents the out-of-phase component.

If an in-phase electrical load is connected to the instrument, it will display a unity or 100% power factor. The resistor-coil path in this ease would develop its full torque. With no out-of-phase component available, the inductor coil would develop zero torque. The resulting deflection would be full scale, or 1.0, indicating a unity-power-factor condition. Should that applied load begin to shift, causing an out-of-phase condition to occur, the inductor-coil arrangement would begin to develop some torque. This action would be in opposition to the torque developed by the resistor-coil arrangement. As a result of this condition, the total torque of the moving coil would be reduced somewhat, thus causing something less than an indication of 1. The indicating scale of a power-factor meter is in decimal values from 0 to 1 or in percentages from 0 to 100%.

Electrical Energy Measurement

The amount of electrical energy that a building uses during operation is constantly monitored by a watt-hour meter. This instrument responds to the amount of voltage and current that is delivered to the system to do work. Since power is expressed as voltage times current, we can describe electric energy as amount of power used for a given period of time. Watt-hours (Wh) and kilowatt hours (kWh) are common measures of electrical energy. The prefix "kilo" of one term is a metric expression denoting 1000. The electrical utility company determines power consumption by kilowatt hour values.

A kilowatt hour meter is a rather unusual electrical instrument that performs several key functions. It senses the value of applied voltage and current, measures the elapsed time it is being used, multiplies these quantities together, and records the amount of power consumed on circular dials. Meters of this type are permanently attached to the service-entrance equipment of the electrical power system. A tabulation of the total power being consumed by the system is permanently registered by the dial-hand location on the meter. Periodically, at 1- or 2-month intervals, the dials are read to determine the power consumption for a given period.

A dial type of kilowatt hour meter is shown in Figure 10-36. To determine the kWh value indicated by this meter, begin with the first dial at the left. The value indicated by this dial is the number that the

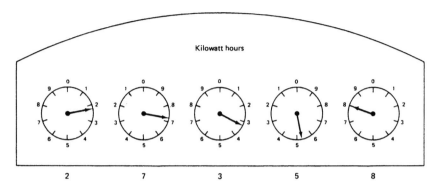

Figure 10-36. Dial-type kilowatt hour meter.

pointer has just passed while moving in the clockwise direction. This number 2 represents the most significant value of the total reading. The next dial (7) indicates the second most significant number value, with the dial turning counterclockwise. The third dial from the left indicates a 3 while turning clockwise. The fourth dial, while turning counterclockwise, indicates a 5. The fifth dial shows the least significant value (8) while turning clockwise. The total value indicated by the meter is 27,358 kWh.

A kWh meter reading does not have meaning when viewed as a single value. The reading does become meaningful, however, when compared with a value taken from the same meter at a different time. As an example, suppose that the previous kWh reading was taken on October 31. On September 30, the same meter recorded a value of 26,923. The amount of electric power consumed during the month of October is therefore equal to 27,358 − 26,923, or 435 kWh. This number times the cost per kWh of electricity indicates the amount of the electrical utility charge for the designated billing period.

The operation of a kilowatt hour meter is similar in many respects to that of a conventional wattmeter. A voltage-sensing coil connected across the incoming power line is used to monitor the applied line voltage. Similarly, an independent current-sensing coil is connected in series with one side of the power line to monitor current values. Both of these coils are mounted inside the meter enclosure and attached to a metal frame as shown in Figure 10-37. Voltage and current values produced by the power system cause an electromagnetic field to appear

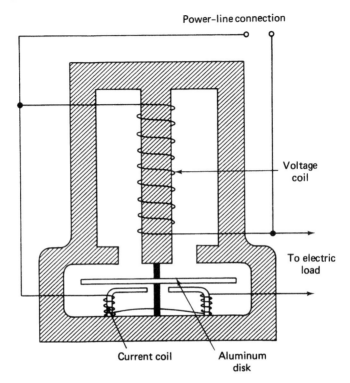

Power–line connection

Voltage
coil

To electric
load

Current coil   Aluminum
disk

Figure 10-37. Kilowatt hour-meter construction.

around each coil. This action causes the aluminum disk to rotate like a motor. The rotational speed of the disk is related directly to the amount of power consumption. A gear train driven by the disk shaft turns the dial mechanism. Power consumption recorded by the instrument is based upon indicating the hand location of the respective dials.

Electrical energy may also be measured by digital instruments. This type of kWh meter is all electronic and has no moving parts. It is permanently attached to the service-entrance equipment and derives its operational energy from the source under test. Sensed voltage and current values are applied directly to analog-to-digital converters. The resulting output values are then multiplied together, counted, and applied to the digital display. Cumulative output indications may displayed for one billing period and then be reset, or may run continuously according the needs of the utility company. Figure 10-38 shows a representative digital kWh instrument.

Figure 10-38. Digital kilowatt hour instrument. (*Courtesy of Energy Research Associaates.*)

Power-Demand Meters

Power-demand meters are designed to measure the peak power (kW) used as compared with the average power (kW) being used. This ratio of power consumption is an indication of the amount of power that the utility company must supply above its predicted average value for a particular building. In many areas of the country, costly penalties are imposed upon a consumer whose peak power demand exceeds the average power demand. A more consistent utilization of power can be realized if a power-demand meter is monitored.

The reading of a kilowatt hour demand meter is similar in many cases to that of the kilowatt hour power meter. A common practice today is to build the demand meter into the kWh power meter. Figure 10-39 shows two rather common types of demand meters.

The multiple-dial demand meter of Figure 10-39 has two complete dial mechanisms. The top four dials of the instrument perform the kilowatt hour power meter function described in the electrical energy measurement section. The lower three dials serve as the power-demand indicators. The demand value is read in the same way as the kWh meter. The vertical line between the second and third dials locates where the decimal point appears in a value. Demand readings in some instruments may involve multiplying the dial reading by a number. The multiplication factor, when used, is printed on the face of the instrument. If a factor is not printed on the dial, the demand reading is

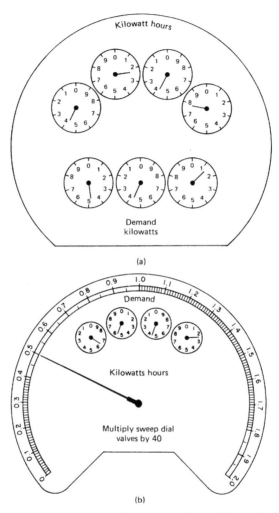

Figure 10-39. Types of demand meters: (a) multiple dial; (b) sweep dial.

indicated by dial values. The demand value of the meter must be reset to zero after obtaining a value reading for a particular billing period. The demand factor applies to only one billing period at a time.

The sweep-dial demand meter of Figure 10-39 is designed to record the total kWh power consumption. The sweep-hand dial pointer of the instrument will point and hold to the maximum kW demand value for a given billing period. In practice, the sweep-hand reading is multiplied by a number printed on the meter face to obtain the demand value. The sweep hand must be reset to zero by the reader when starting a new billing period.

Power-demand meters are available in digital models as well. Unlike the analog styles discussed previously in this chapter, the digital models have an LDC display for showing the power consumed. Additional features of the digital meters include data transmission capability that allows the meter to be read from remote locations. This feature is for purposes such as billing by an energy provider.

FLOW-MEASURING INSTRUMENTATION

Nearly all building operation is influenced in some way by flow instrumentation. This ranges from the measurement of liquid fuel, natural and bottled gas, to water-metering equipment. A person responsible for building operation should have some understanding of flow instruments in order to recognize malfunctions and evaluate system consumption.

Flow instrumentation that is of major concern in building system operation is classified as totalized flow measurement. Instruments of this type are designed primarily to respond to the total amount of flow that moves past a given point during a specified period of time. The measurement of liquid fuel, gas, and water all respond to some type of totalized flow instrumentation.

Liquid-flow Measurement

The measurement of liquid fuel and water is primarily achieved by the same basic type of instrument. Operation is based upon the positive-displacement principle. Known amounts of liquid flowing through a pipe are simply divided into parts and counted. Totalization is achieved by recording each part on an indicating dial or meter. Most instruments of this type do not employ an outside form of energy. Operational energy is extracted from the following force of the liquid stream passing through the instrument.

The nutating-disk flowmeter of Figure 10-40 is a very popular liquid totalizing instrument. The moving part of this meter is used to

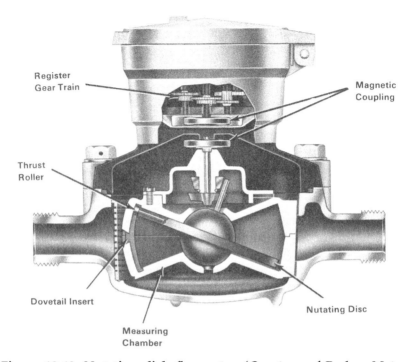

Register Gear Train

Magnetic Coupling

Thrust Roller

Dovetail Insert

Nutating Disc

Measuring Chamber

Figure 10-40. Nutating disk flowmeter. (*Courtesy of Badger Meter, Inc.*)

separate the applied fluid into discrete measurable parts. It consists of a radially slotted disk with a cup-shaped ball bearing at the bottom and a pin extending from the top. This assembly is positioned so that it divides the meter into four chambers. Two of these are located above and below the disk on the input side of the instrument. The alternate two chamber positions are transposed and located on the outlet side of the meter.

Liquid flowing into the metering chamber causes the disk to rotate in a wobbling manner. Each revolution or nutation of the disk causes a given volume of liquid to pass through the instrument. The free-moving shaft on the top of the disk moves in a circular motion with the flow of liquid through the chambers. This action is used to drive a gear train that registers each rotation. A counter attached to the mechanism records and totalizes and amount of flowing liquid. The accuracy of a nutating-disk flowmeter is ±1% for most building installations.

Water Meter

The amount of water supplied to a building during an operational period is permanently recorded on a positive-displacement flowmeter. The nutating disk type of instrument is widely used today for residential and commercial installations. Water quantities are recorded in cubic feet (ft^3), cubic meters (m^3), or gallons. The unit of measure is indicated on the face of the meter.

A representative direct-reading water meter is shown in Fig 10-41. The circular dial of this meter is used to indicate the instantaneous flow rate of the system and is not part of the value. Consumed water values are tabulated on the direct-reading cyclometer dial. To determine the amount of water consumed for a given billing period read the tabulated value indication. In this situation the value is 0003621 ft^3. A previous reading of 0002754 ft^3 was recorded at the beginning of the billing period. Consumption during the period is therefore 0003621 — 0002754, or 867 ft3. Conversion of cubic feet to gallons is achieved by multiplying the cubic feet value by 7.48. In the example, 867 ft^3 × 7.48 = 6485.16 gallons consumed during the period. In some systems, billing is based upon gallons consumed whereas others are determined by ft^3. At present metric values are rarely used in water meters.

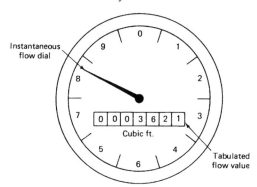

Figure 10-41. Direct-reading water mater.

Gas Flowmeters

The positive-displacement principle of flow measurement is not very suitable for gas instrumentation. Gas flow, for example, must be continuous without a change in pressure. Measurement must not generate pulsing or any form of flow interruption. In addition to this, flow must not be adversely influenced by changes in pressure or temperature. To solve these problems gas-flow instruments are usually of the bellows type of construction.

A typical dry type of bellows flowmeter is shown in Figure 10-42. The operation of this instrument is based upon the transfer of gas between four sealed chambers. A flexible gas-tight bellows is located in two of the chambers. The instrument is designed so that gas entering the meter flows alternately in and out of each chamber compartment by the action of a sliding valve. This action is used to drive a crank mechanism that controls the flow gas alternating between each bellows. Alternate displacement of gas between each chamber by bellows switching causes the output to have a continuous flow.

Figure 10-42. Bellows-type gas flowmeter. (*Courtesy of American Meter Company.***)**

A crank mechanism attached to each bellows is used to drive a gear train. The transfer of gas alternately between the input bellows/chamber and the output bellows/chamber is counted and registered on the indicator. Totalized flow is measured in cubic feet (ft^3) or cubic meters (m^3). Instrument accuracy is ±0.5% with a total flow of 10^7 ft^3 for a period of up to 5 years or greater.

Two representative direct-reading meters are shown in Figure 10-43. The clock-like indicator of Figure 10-43(a) shows totalized flow on the top four dials. The 1/4-foot and 1-foot dials are used to test meter

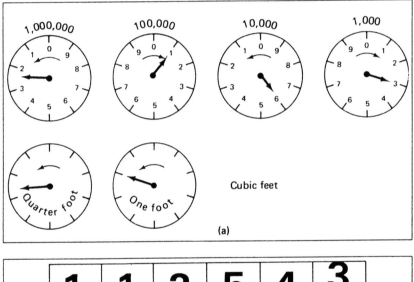

(a)

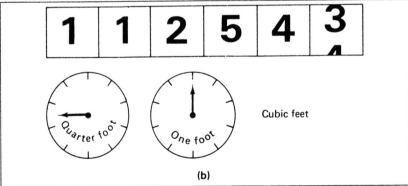

(b)

Figure 10-43. Types of gas meter indicators: (a) clock dial; (b) cyclometer.

operation and to evaluate consumption under controlled usage conditions. The capacity value above each dial indicates the amount of gas consumption required to produce one revolution of the respective dial. Typical values are 1,000,000 ft^3, 100,000 ft^3, 10,000 ft^3, and 1000 ft^3.

To determine the amount of gas consumed during a billing period, two readings are needed. Assume that the 2,163,000 dial reading of Figure 10-43(a) was taken at the end of the billing period. A reading taken at the beginning of the billing period indicated 2,152,000 ft^3. The total gas consumption for the billing period is 2,163,000 – 2,152,000, or 11,000 ft^3. This value, multiplied by the cost of a cubic foot of gas in a particular locality, indicates the amount to be paid for gas usage.

A cyclometer type of gas indicator is shown in Figure 10-43(b). The six-place numbered indicator is read directly in cubic feet of gas. The quarter-foot and one-foot dials are used to show meter operation and to monitor small quantities of gas consumption under controlled conditions. Cyclometer indicators are replacing the clock-like dial meters in most installations today. Readings are taken at the beginning and ending of a billing period to determine total gas consumption. Billing for large consumers of gas may be indicated in hundred cubic feet (ccf) or thousand cubic feet (Mcf) values.

Chapter 11

Energy Management/
For Going Green

INTRODUCTION

Earlier in the book, we stated that energy management is a continuous planning process that is used to accomplish the efficient use of energy in a building. Energy management for one or more buildings requires business decisions based on initial cost of new equipment, cost of maintaining that equipment, and payback time for the investment.

The main thrust of energy management is fairly simple. Its major objective is to conserve energy which in turn saves money. To have effective energy management, a person must have energy awareness and the desire to save both energy and money.

Obviously, it will not always be economically practical to add insulation, change lighting systems, or to replace inefficient heating or air-conditioning systems. However, there are a number of ways that these points can be supplemented at a modest cost. When these are combined with reductions already in place, the energy conservation goals can still be met. Remember that a properly maintained and managed building, on the average, will consume 20% to 30% less energy than a poorly maintained and managed one. If this 20% to 30% reduction in energy is spread to every building, the foreign oil demand will be cut in half, savings would be substantial, and environmental concerns would be reduced.

ENERGY USE IN BUILDINGS

The use of energy in buildings can be viewed in terms of the amount of fuel or type of energy purchased. The cost of energy used in a building is calculated according to a rate charged by the supplier plus the measured quantity of energy used by a system. Typical quantities include gallons of oil, cubic feet of gas, pounds of steam, and kilowatts of electrical energy. These amounts are used to calculate monthly energy bills for a building. It is a common practice to look at energy in terms of Btu per square foot of building area. Each energy quantity can be converted to an equivalent amount of heat. For example, 1 kilowatt hour of electrical energy is equivalent to about 3413 Btu of heat. These Btu conversions, shown in Table 11-1, are used to determine Btu per square foot of building area.

Table 11-1

Energy source	Heat equivalent
Electricity	3413 Btu/kilowatt hour
Natural gas	1000 Btu/cubic foot
Fuel oil (No. 2)	138,000 Btu/gallon
Coal	26 million Btu/short ton
Steam	900 Btu/pound
Propane	21,500 Btu/pound

Simple Payback Period

A term often used in regard to energy conservation is payback period. Payback period can be estimated to determine the feasibility of making a capital investment that will provide energy savings. Simple payback period is an indication of the amount of time it will take for an organization to recover the cost of an energy-saving investment which is purchased. The formula for estimating simple payback period is

$$\text{payback period} = \frac{\text{cost of improvement (\$)}}{\text{average yearly savings (\$)}}$$

For example, if adding insulation to a ceiling at a cost of $2000 is expected to provide a $1000 savings each year, the simple payback period would be $2000/$1000 = 2 years. Thus, the improvement would pay for itself in a 2-year period. Simple payback does not take into account escalating fuel costs, interest on capital, and other financial factors. However, it does provide a relative index of the economics of potential energy conservation measures that could result in energy and money savings.

Return on Investment (ROI) through Energy Management

Managers in most business organizations are familiar with the concept of return on investment (ROI). Many homeowners have had to become familiar with the term when an improvement, that is other than superficial, is made. It is used by both groups to analyze their investment. In years past, most individuals have considered energy conservation as insignificant in the total financial picture. Inflated energy costs have made ROI for energy-use reduction more favorable. Energy costs should now be a major concern for any person or organization. Many building modifications for energy conservation brought about by energy management programs have an excellent return on investment.

Retrofit of Buildings

When a building is modified in some way which involves construction or additional equipment, this is referred to as retrofit. Retrofit for the purpose of energy conservation could involve such modifications as redesign of a lighting system, changes in HVAC systems, adding insulation to roofs or walls, or adding equipment to reduce electrical power demand.

In considering retrofit of existing buildings, owners and operators should consider making several changes at one time rather than small changes made over a long period of time. An energy audit of a building can assess ways in which retrofit will be valuable. Several methods of saving energy and saving money can be discovered through a careful energy audit. These methods should be coordinated in such a way to have the greatest impact on energy conservation and thus the most desirable ROI. When we consider the escalation of energy costs, we see that the ROI for energy conservation is very favorable. If energy costs increase 30%, so do annual savings, and the ROI increases substantially. With this in mind, retrofit of a building should be more desirable if initiated quickly to minimize inflation of material and equipment costs. Retrofit of existing buildings with energy conservation in mind should be a valuable financial investment.

CONSIDERATIONS FOR EFFECTIVE
ENERGY MANAGEMENT

There are certain factors that should be considered for effective energy management. Some of these are energy-efficiency standards and codes which have been developed. Several considerations that affect energy management are discussed in this section. These include the Emergency Building Temperature Restriction (EBTR) program, energy-efficiency ratings of equipment, and the Building Energy Performance Standards (BEPS).

ASHRAE Standards

The American Society of Heating, Refrigeration, and Air-Conditioning Engineers (ASHRAE) has developed certain standards which deal primarily with energy conservation in buildings. ASHRAE standards provide guidelines for energy conservation through proper building design. Technical data that show the effect of various aspects of building design are presented in ASHRAE standards.

ASHRAE Standard 90-75, "Energy Conservation in New Building Design," deals with the major design characteristics of a building which affect energy use. Systems included in ASHRAE 90-75 are (1) the exterior envelope, (2) HVAC systems, (3) HVAC equipment, (4) service water heating, (5) electrical distribution systems (6) lighting systems, and (7) the use of solar, wind, or nondepleting energy sources.

BOCA Basic Energy Conservation Code

The Building Officials Code Administration (BOCA) has developed a code that can be used as a guide for energy conservation in buildings. The BOCA Basic Energy Conservation Code is one of several code books developed by BOCA. These codes are recommended for adoption by state or local governments for building code administration and enforcement.

The BOCA Basic Energy Conservation Code has several sections devoted to various aspects of energy conservation in buildings. Among the sections included are (1) building envelope, (2) warm-air heating, ventilating and air-conditioning systems and equipment, (3) plumbing systems, (4) electrical systems, and (5) alternative systems.

Emergency Building Temperature
Restriction (EBTR) Program

The effort by the federal government to achieve energy conservation in commercial and industrial buildings has resulted in the Emergency Building Temperature Restriction (EBTR) Program. The EBTR Program presently recommends that commercial and industrial buildings maintain room temperatures at 65°F in winter and 78°F in summer for all areas served by a heating or air-conditioning system. There has been considerable opposition to this program because of the variations in HVAC systems and the effects of changing thermostat settings on the energy consumption of systems.

The EBTR program also recommends setting thermostats on domestic hot-water systems at 105°F or less, except where there is a special need for higher-temperature water. Another provision of the EBTR program is that commercial and industrial buildings turn off their air-conditioning system when a building is unoccupied.

Energy Efficiency Ratio (EER)

The energy efficiency ratio (EER) is an index of the efficiency of a system. EERs are presently used to rate the energy use of several types of appliances typically used in homes, such as refrigerators, room air conditioners, freezers, clothes washers, dishwashers, and water heaters. The EER of a system is placed on a label that is attached to the system. The EER labeling of other types of equipment is expected to take place in the future.

Essentially, the higher the EER number is, the more efficient the energy use of the equipment is. Equipment with higher efficiency uses less energy and thus costs less to use. Systems having higher numbers usually cost somewhat more. Usually, additional initial costs can be recovered in a short period through reduced operating costs. The EER label also lists the price of similar models with the lowest and highest energy costs. In addition, estimates of yearly operating costs are included on this label, which is referred to as an "Energy Guide."

Building Energy Performance Standards (BEPS)

As a result of the efforts of the U.S. Department of Energy, Building Energy Performance Standards (BEPS) are being developed.

The standards should provide a reference to use in building design to accomplish energy efficiency. BEPS are, at the present time, in the developmental stages.

DEVELOPING AN
ENERGY MANAGEMENT PROGRAM

It is important for building owners and operators to develop effective energy management programs. Energy management programs, when administered properly, can effectively reduce the amount of energy used in a building. The effort will not only save money but it will conserve our national or world fuel resources.

Energy conservation, by itself, is not energy management. Energy conservation is doing what is easy and economical in the short term. Energy management, in contrast, is the long-term commitment of one or more individuals.

Major objectives of an energy management program for any residential, commercial or industrial building should be:

1. To reduce energy use without making impractical financial investments.

2. To maintain a comfortable living or working environment.

3. To assure that the building meets federal and state regulations pertaining to energy use.

4. To improve the efficiency of equipment and reduce the operating costs.

Many methods of managing and conserving energy have been discussed in this book. The critical part of any energy management program is commitment to the saving of energy by the building manager or owner. These people must be convinced that energy management saves them money and is important for our energy resources. Building managers and owners should always keep energy conservation in mind and develop realistic objectives for energy use. The key element is "awareness."

SUGGESTIONS FOR BUILDING
OWNERS AND OPERATORS

There are several suggestions that building owners and operators should consider when developing an energy management program.

1. Plan the program with care.

2. Hire or contract with an energy specialist to make an analysis of energy use in the building.

3. Delegate someone dependable to supervise the overall energy management effort.

4. Collect and analyze data on fuel and energy cost.

5. Maintain control over the way in which energy is used in the building (develop a "policy" regarding energy use).

6. Hire professional consultants (if it is financially feasible) to analyze energy use in the building and make recommendations for modifications that will save energy.

7. Maintain accurate records of equipment operating schedules and room occupancy.

8. Urge employees to help in the conservation effort by turning off lights, using as little hot water as possible, closing doors, and maintaining proper thermostat settings.

9. Conduct periodic checks to evaluate the effectiveness of the energy management program and suggest ways of improvement.

Energy Management Teamwork

To be effective, energy must be saved by a team. An individual, in reality, cannot do this alone. This person might be helped by another member of the family, a specialist in a particular area of energy conservation, and possibly a contractor. In a larger building, a team might consist of a leader who is either a technician or a member of management, plus a specialist in each system area.

One method of promoting energy management teamwork in an organization is to develop an energy management team which involves managers and employees. The purpose of such a team should be to observe energy use and recognize areas where energy could be saved. The team should recommend changes that would help conserve energy in the building where they work. By involving several people on a team, an awareness of energy conservation should be apparent throughout the organization. Several companies have tried the team approach to energy conservation. Most of these companies have saved on energy cost and thus saved money. A good energy management effort requires detailed planning, organizing, and controlling. This requires a definite commitment from top management to the energy conservation effort.

ENERGY AUDIT

It should be obvious from your reading that most energy management does not just happen. Just one system not functioning properly could foul up an entire home or process. An energy audit of a building assists an individual or team to identify these potential or problem areas.

Audit Preparation

An energy audit is not a difficult thing to do, but in order to do it correctly certain steps must be followed. The first step is to define the project. In other words, is making improvements worth the effort? Secondly, an energy survey needs to be performed. Take a walking tour of the building to get a feel of what needs to be done to meet your goal of energy efficiency. Before touring the area, make certain it is conducted by someone who is familiar with the various mechanical and electrical systems of the building. Making an energy appraisal is the third step. This is done by either using a computer or by hand computation to find the cycles for days, months, and years through utility bills. In other words, find the specific times when there is an energy increase or decrease in the building. This also tells a person the overall energy budget for the building. The fourth step is to identify and develop energy conservation opportunities. This step shows some of the many measures and techniques which can be implemented in a building to reduce consumption. Some areas to be looked at are the building envelope, the lighting, heating, and the cooling systems. If ways are found to make each of these areas efficient then the next step comes into play. The fifth step is the evaluation of these opportunities. The feasibility of these opportunities is then decided. The sixth step is to develop an energy action plan. When the plan is properly developed it will provide documentation of specific energy conservation measures and will permit the operation of the building at the lowest practical energy use and cost. A maintenance plan will also be included in this particular step. The final step, developing an energy audit, will not only provide an assessment of the effectiveness of the energy management plan, but also can give data to make certain these new steps are the most efficient for the building. This step also helps to identify where additional conservation efforts would be useful.

ENERGY AUDIT CHECKLIST

There are several systems of a building that should be checked as part of an energy audit. Some of these areas are described next.

1. Conduct a lighting survey.
    Determine the number and types of lighting fixtures.
    Determine the foot-candles of light in each area.
    Check the switching methods used to turn lights on and off.
    Determine the average hours of light usage in each area.
    Review current IES guidelines for lighting systems.

2. Conduct a building utilization survey.
    Determine how each area of the building is actually used.
    Determine the hours of use of each area.
    Determine the needs for HVAC system use in each area.
    Determine the approximate numbers of people who occupy each area and how these numbers vary over a 24-hour period.

3. Conduct a survey of the building envelope.
   Check condition of walls, roofs, floors, ceilings, and entries.
   Determine if adequate insulation is used in all areas.
   Check areas for air infiltration.

4. Conduct a survey of electrical equipment.
   Check electricity bills for demand charges and determine
   equipment utilization schedule.
   Check transformer ratings and loading.
   Check the power factor of the electrical system.
   Determine loading of all large electric motors.

5. Conduct a survey of steam and hot-water systems.
   Check the general condition of steam and hot-water distribution
   systems.
   Make boiler efficiency tests.
   Determine utilization of equipment on a daily and seasonal basis.
   Check to determine the adequacy of insulation of steam and hot-
   water distribution systems.

6. Conduct a survey of the HVAC systems.
   Determine the most energy-efficient temperatures for system
   operation throughout the heating and cooling seasons.
   Check exhaust air fans, supply- and return-air systems, and
   outdoor air quantities and keep operating data for daily and
   yearly periods.
   Make operational checks to determine if systems are operating at
   maximum efficiency.
   Review ASHRAE Standard 90-75.

7. Conduct a survey of special-purpose energy-consuming systems.
   Check temperature of the hot water in the domestic hot-water
   system (105°F or less is recommended).
   Check special-purpose process equipment utilization.

   Some of these checks may be made rapidly, whereas others require accurate record keeping over a long period of time. However,
each of the general items listed is important for a successful energy
audit of a building.

ENERGY SAVING THROUGH
PREVENTIVE MAINTENANCE

Preventive maintenance performed on equipment can make a significant contribution to energy savings. This is particularly true for electrical and mechanical systems, whose condition affects their operating efficiency. Preventive maintenance (PM) procedures are used to take corrective action on equipment before breakdown occurs. PM should be done on a scheduled basis. In many cases, PM can be computerized for large buildings with a maintenance staff. PM can save companies money because precise scheduling of maintenance activities increases the operating efficiencies of equipment and increases the lifetime of equipment. Increased efficiency of equipment will reduce the amount of energy used.

Computerized PM systems can be used to inventory equipment, list its location, manufacturer, age previous maintenance, and any other pertinent information. Such systems can provide annual equipment inventories very quickly and simply and forecast maintenance needs.

EQUIPMENT SCHEDULING

One obvious method of conserving energy is to *turn equipment off* when it is not being used. This can be done by either manual or automatic methods. Automatic control is the most exact method of accomplishing control in large buildings. Many types of automatic controllers and timers are commercially available to control the scheduled use of equipment.

A scheduling method that can be used effectively to save energy is *night setback*. An automatic timing system for night setback allows loads to be sequentially turned off and then back on, or an entire system (such as lights) to be turned off and on at the same time. Night setback can be used on HVAC systems to change room temperatures and reduce energy use during nonbusiness hours.

Duty cycling is another method used to schedule equipment operation for energy conservation. This method allows loads to be turned on and off at a predetermined rate so that the system is not operated at times when it is not needed. Various types of equipment used in buildings (such as ventilation fans) lend themselves well to duty cycling.

Load leveling is yet another method that may be used to schedule equipment use. This method can be used to reduce electrical demand changes imposed by utilities companies. Demand was discussed in Chapter 8. To reduce peak electrical power demand, a monitoring system, such as a load leveler, can be used to limit the use of major power-consuming equipment at the same time. Since electrical demand charges are very high in many areas, load levelers could have a short payback period. A microprocessor-based load control system is shown in Figure 11-1. In addition to load leveling a system such as the one shown can pro-gram the loads of a building over a daily, weekly, or monthly operational schedule. Several systems are available which use micro-processor control for energy management. An internal view of a microprocessor-based energy management control panel is shown in Figure 11-2.

Thermostat Covers

Most buildings have thermostats located in several locations to control the temperature in specific areas. Employees usually have access to these thermostats; therefore, they can make adjustments in temperature as they desire. Often, changes in thermostat setting adversely affect the energy use in a building.

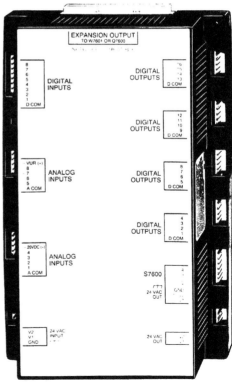

Figure 11-1. Microprocessor-based load control system. (*Courtesy of Honeywell, Inc.*)

Figure 11-2. Internal view of a microprocessor-based energy management system. (*Courtesy of Pacific Technology, Inc.*)

The thermostat cover shown in Figure 11-3 is a simple and inexpensive way to avoid changes in temperature setting. These covers are available with locking systems so that thermostats may be set to the desired temperature by the building manager and maintained in that position. Employees would not be able to make thermostat adjustments because they would not have access to the temperature control.

Cogeneration Systems

The term *cogeneration* refers to the production of energy in more than one form at the same time. A common example of cogeneration is the production of electrical energy and heat energy in the form of steam or hot water simultaneously. The extent to which cogeneration is valuable to energy conservation depends upon many factors.

There are three major areas of use for cogeneration. These are (1) large industrial and commercial buildings which use a large amount of process heat; (2) a producer of process heat for several users, with electrical energy also produced and sold to a local utility company;

Figure 11-3a. Thermostats: "Energygard" thermostat cover. (*Courtesy of Energy Controls Inc.*)

Figure 11-3b. A programmable thermostat with touch-screen.

and (3) utility companies which produce electrical energy and process heat. Most uses of cogeneration provide very efficient utilization of steam. For instance, an industry that uses process steam can also use the steam to rotate a steam turbine. The steam turbine is used as the source of mechanical energy to rotate an electrical generator. Cogeneration systems could be valuable for many organizations to consider as a means of conserving energy.

Water-Pressure-Reduction Valves

A method of saving energy consumed through water usage is the installation of water-pressure-reduction valves. The water pressure used in many areas far exceeds the amount actually needed. The primary savings of this method would be in actual water consumption. A reduction in hot water used would decrease the amount of energy required for the domestic hot-water system. Manufacturers of these devices also claim that reduced pressure would curb maintenance cost of associated equipment to some extent. A water-pressure-reduction valve used for a shower hood is shown in Figure 11-4.

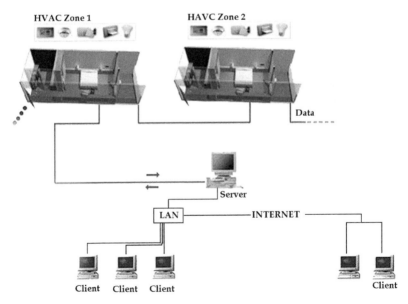

Figure 11-4. A flow diagram of an internet-ready computerized energy management system.

COMPUTERIZED
ENERGY MANAGEMENT SYSTEMS

Several types of highly sophisticated computerized energy management systems are available today. These systems may be used effectively to control the environment of a building with a control computer system. Such systems constantly monitor and adjust the operation of systems in a building which consume energy. They may be used to control the operational times of fans, pumps, and motors. Also, they are capable of performing such functions as demand limiting or load shedding, boiler and chiller operation control, equipment operational scheduling, ventilation system control, system maintenance, and security and safety functions.

One simple application of a computer or computer control to an existing system is the addition of a programmable thermostat. By using solid-state devices to measure temperature, a programmable thermostat can execute a program from an internal memory to alter the temperature in the zone it controls by time of day or the day of the week. Typically, the thermostat is programmed by schedules. Daily schedules include a temperature setting for a time to wake, another temperature for a time while away at work, a third setting for return from work, and last a temperature setting for sleep time. Depending on the complexity of the thermostat, these schedules can be set for each day of the week, altering the schedule on weekends if desired. Because it is microprocessor based, the programmable thermostat can incorporate other features such as a vacation setting, lockout of unauthorized users, and automatic changeover between seasons. Also, some programmable thermostats act as data recorders tracking the hours the system has ran, temperature extremes, and maintenance timers such as filter replacement. An example of a programmable thermostat with a touch-screen programming feature is shown in Figure 11-3.

Many computerized energy management systems are available. Each system has its own unique features. Figures 11-5 through 11-9 are illustrations of several types of energy management systems which are on the market today. It should be pointed out that many other manufacturers make energy management systems that are used to reduce energy use in buildings.

There are several factors that an organization should consider before purchasing any type of energy management system. Computerized energy management systems are expensive so the payback period should be carefully calculated before a commitment to purchase is made, alternative energy control methods should be considered. These alternatives include timers and programmable controllers. The complexity of programming the system should also be considered. Possibly, a full-time or part-time operator will be needed to program the system and make operational checks and adjustments. In addition, installation and maintenance costs must be considered. Once these considerations have been carefully weighed, the proper type of energy management system should be specified.

Programmable Controllers

Several companies manufacture programmable controllers. They can be used effectively to control the operation of energy-consuming equipment, although their primary function has been for industrial process control. Essentially, a programmable controller is a solid-state system that can be used to control the operational time of a

Figure 11-5. An energy management control center.

Figure 11-6. Square D Energy Management System. (*Courtesy of Square D Co.*)

Figure 11-7. Independent Energy's Energy Manager (for solar heating systems). (*Courtesy of Independent Energy Inc.*)

Figure 11-8. Johnson Control's Zone Terminal. (*Courtesy of Johnson Controls, Inc.*)

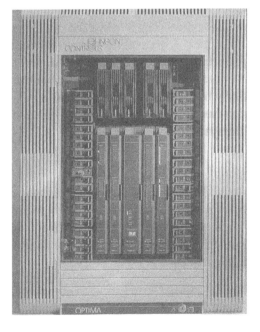

Figure 11-9. Johnson Control's Intelligent Lighting Controller. (*Courtesy of Johnson Controls, Inc.*)

system connected to it. They may be used to turn loads on and off and to control the sequencing of equipment, such as for load shedding. A programmable controller could be used to control many different pieces of equipment from a central location. These systems differ from computer control systems in that they are ordinarily easier to operate and easier to modify functions performed. They can also be adapted easily to various types of environments, particularly in industry. A programmable controller is shown in Figure 11-10. Figure 11-11 shows how it controls the cooling system of a building.

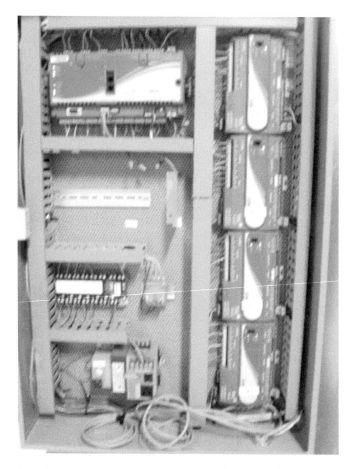

Figure 11-10. Inside a programmable control cabinet for a commercial air handler.

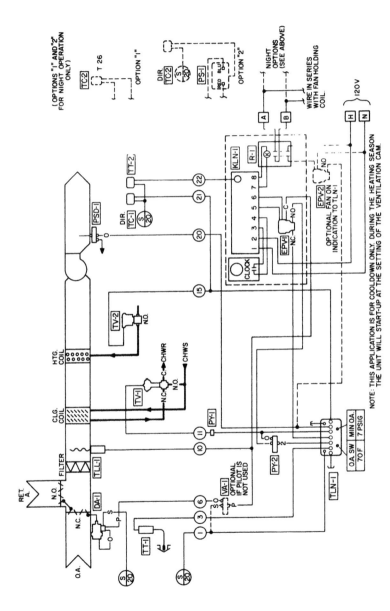

Figure 11-11. Schematic of a programmable controller used to control a cooling system. (*Courtesy of Johnson Controls-Systems and Services Division.*)

Timers

It is possible to use timers to accomplish the desired type of energy control. Timers may be used to control the operational time of many types of equipment used in buildings. Lighting, heating, and cooling systems, as well as individual machines, can be controlled by on/off timers such as the one shown in Figure 11-12.

Ice Bank Systems for Air Conditioning

Ice bank systems for air conditioning, such as the one shown in Figure 11-13, are designed to help conserve energy. The operational principle allows the air-conditioning compressor to form ice during low-demand periods. The ice forms around evaporator coils which are immersed in a storage tank. During high-demand times, when cooling is needed, chilled water is circulated through the air-handling system. Figure 11-14 shows an ice bank storage tank and evaporator, and Figure 11-15 shows a typical installation.

Pressure Controllers

Systems that operate on a pressure basis ordinarily use more energy when they produce more pressure. Steam boilers are a typical example of pressurized systems. The pressure controller shown in Figure 11-15 may be used in conjunction with a timer to set boiler pressure back automatically during low-demand times and then increase the pressure during high-demand times.

COMPUTER NETWORKED CONTROLS

Many of the previous controllers described in this chapter could be described as dedicated types of control. They are capable of taking in data from var-

Figure 11-12. Timer control. (*Courtesy of Honeywell, Inc.*)

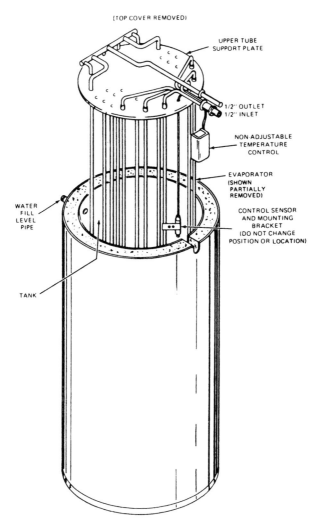

(TOP COVER REMOVED)

UPPER TUBE
SUPPORT PLATE

1/2" OUTLET
1/2" INLET

NON ADJUSTABLE
TEMPERATURE
CONTROL

EVAPORATOR
(SHOWN
PARTIALLY
REMOVED)

WATER
FILL
LEVEL
PIPE

CONTROL SENSOR
AND MOUNTING
BRACKET
(DO NOT CHANGE
POSITION OR LOCATION)

TANK

Figure 11-13. Ice bank storage tank and evaporator. (*Courtesy of A.O. Smith Corp., Consumer Products Div.*)

ious sensors, running that data through software, and changing the outputs of the system based on the software or programming. All this could be accomplished with little to no human intervention, except a user's console. Traditionally any changes in these systems would have to be done on-site, near the physical location of the controller.

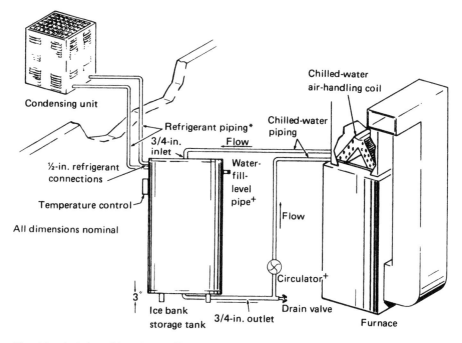

*Should include liquid-line dryer—filter and suction line oil separator.
+Circular should be sized for air-handling coil and have flow-rate-adjustment provisions.

+Opening capped in single tank system. Connect to water-level equalizer line in multiple-tank system.

Figure 11-14. Typical installation of an ice bank storage tank. (*Courtesy of A.O. Smith Co., Consumer Products Div.*)

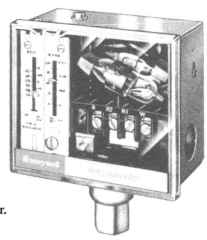

Figure 11-15. Pressure controller.
(*Courtesy of Honeywell, Inc.*)

This certainly limits the use of such controls. Also, it is common for controls in a heating or air-conditioning system to be proprietary, meaning only sensors, actuators, software, and other pieces of an automated system would work together only if offered by the same manufacturer. Because of this proprietary nature, controls and actuators from one manufacturer were incompatible with and could not be used with a system from a competing controls manufacturer. With a need to offer remote control of automation networks and in an effort to reduce the number of proprietary systems, a new generation in the automation of heating and air conditioning began. This evolution started in roughly 1995 when standards and protocols were developed for building automation systems. They are titled BACnet® and LonWorks™. Since 1995, many manufacturers have broken from the tradition of proprietary systems to provide connectivity to one of these network standards.

In order to operate on computer networks, each system needs to be compatible with existing standards for computer networking. TCP/IP, which stands for Transmission Control Protocol / Internet Protocol is one of the standard protocols for communication via the Internet. Also, consideration must be made for formatting that is compatible with the World Wide Web (WWW). Any automation system that desires remote connectivity must be compatible with the languages that format WWW page layout, tags, links, and other types of tools available to authors and designers of WWW pages. These standards exist to ensure text and graphics can exist on a page that is viewable in a web browser. The first language was HTML, HyperText Markup Language, and most web pages are now authored with XML, or Extensible Markup Language. Other computer network considerations include the assigning of a unique address within a network, error checking, and ultimately a compatibility with existing Ethernet architecture for data transfer. Without compatibility with such standards, no HVAC automation system will be compatible with computer networks of today.

BACnet® was developed by ASHRAE, the American Society of Heating, Refrigeration, and Air-Conditioning Engineers. Through a committee of individuals and manufacturers, it was celebrated as the standard to best meet the needs of the entire HVAC industry. Unfortunately there have been setbacks. Aside from some devices that have compatibility, most of the problems plaguing BACnet are apparent problems working with existing computer networking standards. Because it was designed by a committee, has been slow in changing with the fast-paced computer networking industry. This makes for slow growth in an industry known for its growth.

LonWorks™ was developed by Echelon Corporation, and is touted as an open-protocol network. Because this standard is developed by a single company, it has not been prone to the lack of compatibility with protocols that have plagued BACnet. It is much more compatible with computer networking standards, and the web. However, LonWorks devices may not be truly open-protocol network. Each device on the network must have a Neuron chip from Echelon. Also, LonWorks is considered a part of BACnet, but the networks are rarely combined.

As the number of installations increase, various sensors and actuators are added to manufacturer's lists of available products, and the integration with existing computer networks progress, it will be interesting to note if one, both, or neither of these standard automation protocols continue to exist. Also, computer network attacks, or 'hacks' are of an increasing concern in the automation of HVAC. Because the same system that controls the air temperature may also control building security, the concern about this threat is very real. Efforts to make these control networks safe include standard computer networking measures such as firewalls, security policies for network administrators and users, as well as software that is capable of detecting possible intrusions. Home users should begin to take note of these systems as well. While the primary application is aimed toward the commercial user (schools, public buildings, hotels, etc.), these technologies are planned for the home as well, at a time when it is considered cost-effective.

ENERGY CONSERVATION CHECKLIST
FOR ENERGY MANAGEMENT SYSTEMS

Table 11-2 presents a checklist for energy management systems.

Table 11-2: Checklist for Energy Management Systems/For Going Green

Items to check	Corrective action
Building room temperature during occupied periods. Accurately check the temperatures of each room in the building to see if they are too high in the winter or too low in the summer. Keep records of the temperatures maintained in each room. (It is estimated that on the average, a 1-degree change in temperature will cause a 2% change in energy use.)	Adjust thermostat settings to proper level for energy conservation (recommended 68°F for heating and 78°F for cooling); install covers or locks to prevent adjustment of thermostats by unauthorized personnel; replace old thermostats with ones that have a limited range of an adjustment for heating and cooling seasons; install computerized energy management system to control building temperatures.
Building temperature during unoccupied periods. Determine actual hours when the building is occupied; check to see if building temperatures are changed to allow energy conservation during unoccupied periods.	Decrease temperature setting by at least 10°F in winter and turn off cooling system at night and on weekends and holidays when the building is not occupied; install timers or computerized energy management systems to control the times when the heating and cooling systems operate.
Storage rooms and unoccupied spaces. Check to see if storerooms or other unoccupied areas in the building which are ordinarily heated or cooled; see if doors from heated or cooled areas to unheated or uncooled	Adjust the temperature of unoccupied spaces to a proper level in the heating season (55°F recommended); turn off the heat in areas where there is no need to keep items from freezing; use portable heaters

(Continued)

Items to check	Corrective action
areas are allowed to stay open for extended periods of time.	in large areas where there are few people working; turn off the cooling system to all unoccupied spaces and storage areas; place automatic door closers on all doors between areas that are heated or cooled and areas that are not; install a computerized energy management system.
Building activity schedule. Check the schedule of activities for the building to see if there are rooms used unnecessarily during off-hours or for intermittent activity during work hours.	Attempt to schedule activities in the building during off-hours only as necessary; and when activities are necessary, schedule them so that only minimum floor space must be heated or cooled; schedule building maintenance activities to take place during regular hours when possible; install a computerized energy management system to control the temperatures in all parts of the building according to the occupancy.
Heating and cooling system operation. Check to see if the heating or cooling system is operated continuously.	Turn on the heating or cooling system less than 1 hour before personnel arrive at the building and turn it off before personnel leave the building (try different startup and turn-off times until minimum energy usage is reached); install a timing system or computerized energy management system to reduce heating time during unoccupied periods and turn the cooling system off at night.

(Continued)

Table 11-2: Checklist for Energy Management Systems (Continued)

Items to check	Corrective action
"Energy awareness" of employees. Determine whether or not personnel who use the building are aware of the need to conserve energy and proper methods to accomplish conservation.	Make sure that all personnel contribute to the overall energy conservation effort by using written or verbal communication (use signs or reminders when and where appropriate).
Thermostat settings. Accurately measure room temperatures and compare with thermostat settings to see if thermostats work accurately and are positioned in a location to accomplish accurate control.	When necessary, calibrate existing thermostats to accomplish accurate control; replace thermostats when necessary with a type that offers limited range control during heating and cooling seasons.
Relative humidity of rooms. Check the relative humidity of rooms to see if the proper levels are maintained during summer and winter seasons.	Adjust existing humidification equipment accordingly; install new humidification or dehumidification equipment to make conditioning of the air in the building more efficient.
Hot-water temperature. Check the temperature of domestic hot water to see if the temperature is too high.	Adjust temperature of water to proper level (55°F recommended); if the building uses dishwashers, for which high-temperature water is needed, consider installing a booster heater for the dishwasher.
Hot-water use. Check to see if hot water is conserved to the maximum extent; check for drips or leaks in the system.	Install flow restrictors where practical; remove standard water faucets and replace with a self-closing type that has a flow restrictor; repair all leaky pipes, faucets, pumps, or storage containers.

Chapter 12

Alternative Energy Systems/ More Green Possibilities

INTRODUCTION

Several methods of producing electrical power are in limited use at the present time. Some these types of alternative energy systems show promise as a possible electrical power production method of the future. Some of these are geothermal, wind, tidal, biomass and tidal power. We have already discussed solar power at some length in Chapter 9 and have shown the potential for that particular type of energy source. Solar power produces electricity through the use of photovoltaic cells.

This discussion is included to stimulate thought about potential alternative systems. Each of the systems discussed have many potential problems. No matter how serious, experimentation must be conducted to assure that electrical power can be produced economically. Our technology is dependent on low-cost electrical power.

GEOTHERMAL POWER

Geothermal systems have promise as a future energy source for many parts of the world. About 20 miles below the earth's crust is a molten mass of liquid and gaseous matter called magma. When the magma comes close to the earth's surface, possibly through a rupture, a volcano could be formed and erupt. Magma could also cause steam vents, like the "Geysers" area of California. These are naturally occurring vents which permit the escape of the steam that is formed by the water which comes in contact with the underground magma. A basic geothermal power system is shown in Figure 12-1 and the understructure is illustrated in Figure 12-2.

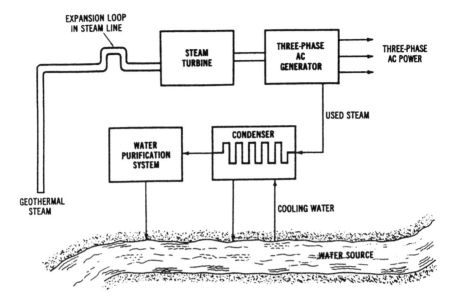

Figure 12-1. Drawing of a basic geothermal power system.

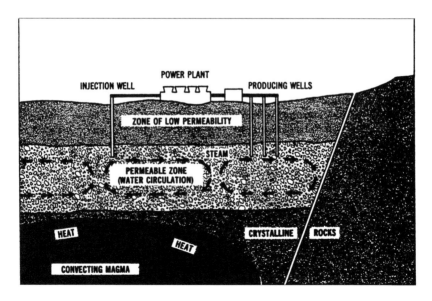

Figure 12-2. Drawing of a basic geothermal power system that illustrates the underground structure.

In the 1920s, an attempt was made to use the geyser named "Old Faithful" as a power source. The pipelines used to harness the geothermal steam would rust because of the steam and the impurities in the gas. Later in the 1950s, stainless steel alloys were developed that could withstand the steam and its impurities. The Pacific, Gas, and Electric Company started development of a power system to use the heat from within the earth as an energy source. The first generating unit at the Geysers power plant began operation in 1960. At present, over 500 megawatts of electrical power is available from the generating units in the Geyser area. The Geyser geothermal power plant is shown in Figures 12-3 and 12-4.

In a geothermal system, steam enters a path (through a pipeline or a vent) to the surface of the earth. The pipelines which carry the steam are constructed with large expansion loops that cause small pieces of rock to be left in the loops. This system of loops avoids damage to the steam turbine blades. After the steam goes through the turbine, it goes to a condenser where it is then combined with cooler water. This water is pumped to cooling towers where the water temperature is reduced. This part of the geothermal system is similar to conventional steam systems.

The geothermal power production method offers another alternative of producing electrical power. This method makes it possible to control energy in the form of steam, or heated water, which is produced by natural geysers or underground channels. The high-pressure steam for power production is made without burning any fossil

Figure 12-3. The Geysers geothermal power plant.

Figure 12-4. An operational geothermal power system.

fuels. This method can be used to drive turbine-generator systems such as the one at the Geysers system in California.

The installation of a geothermal power plant requires the drilling of holes deep into the surface of the earth. One hole may be used to send cold water down into the tremendously hot material under the surface. A hole drilled adjacent to the cold-water hole could be used to bring steam back to the surface. This method is capable of being used in any area of the world, but it requires the drilling of holes up to 10 miles in depth. Otherwise, natural geothermal steam production is limited to active volcanic or geyser areas. Since no fuel is burned, geothermal methods represent one way of saving the valuable and diminishing fossil fuels.

Geothermal for Residential and Commercial Applications

Because geothermal energy is a nearly unlimited resource, it is becoming increasingly popular to utilize this resource as a source of heating and cooling for residential and commercial areas. Touting a savings of up to 60% over conventional heating and air conditioning products with clean and safe operation, many contractors and owners have found geothermal to be a cost effective solution especially during new construction.

The system works on two principles. The first principle is that the temperature underground does not fluctuate with the change of seasons. The second is the theory of a heat pump, the ability to transfer heat from one area to another using pumps and refrigerant. The systems are combined by placing one coil for the heat pump in the dwelling and the second coil buried in the earth or large body of water. In winter, heat is drawn from the second coil in the ground and transferred to the dwelling coil to provide heating. Conversely, in summer, the heat extracted from the dwelling coil is sent to the coil in the earth.

The coils placed in the earth are typically referred to as 'earth loops'. It is the function of the earth loop to effectively transfer the heat to and from the ground. Earth loops have to major classifications, open and closed loops. Open loops are those that utilize groundwater as a heat source, typically pumped from a well. More popular are the closed loop systems where coils of pipe are submerged in a lake or pond or buried in the earth. In such closed loop systems, a media such as ethyl glycol (antifreeze) is used to transfer the heat. There are four methodologies for installing such loops. (1)Pond or lake loops are closed loops installed at the bottom of a body of water. (2) Vertical loops are closed loops that use small pipes buried in deep wells, up to 400 feet. (3)Horizontal loops are closed loop pipe placed in trenches and (4) Open loops are those that use groundwater. The selection of the best loop type is made by the installer and typically depends on the availability of groundwater and land area. A fifth technique for placing the earth loops is increasing in popularity. Rather than install the pipe flat in a horizontal loop, the pipe is coiled in a 'slinky' configuration. This allows much more surface area of pipe to be exposed to the earth.

Geothermal systems are gaining in popularity for many reasons. They have low operating expenses, lower maintenance costs, and are esthetically appealing compared to conventional systems. The low maintenance costs can be directly attributed to the simplicity of the system. The earth loop tubing typically has a long warranty and it is estimated that the actual tubes can last over 50 years. System leaks are rare, and there is no external condensing unit, flue, pipes, or such items to detract from the exterior of the building. In addition, there are no concerns of carbon monoxide as with fossil fuel heating. Figure 12-5 shows a typical installation of a residential geothermal unit. Note the simplicity of the system, the only item not shown is the earth loop.

Such a system operates inexpensively because the energy in the earth is essentially 'free' and the only cost of running the system is the energy to run the pump and air handler plus minor maintenance. Also, the heat removed can easily be piped to a water heater to further reduce costs for hot water. This configuration is sometimes referred to as desuperheater and typically works during the cooling cycle. To

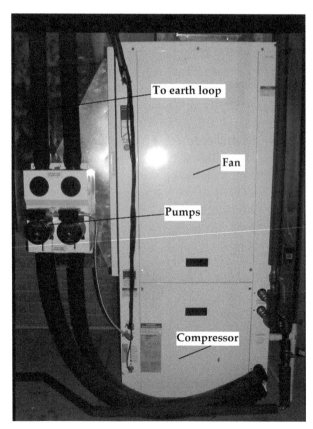

Figure 12-5. Steam expansion loops of a geothermal system.

achieve heating and cooling by using geothermal energy, no fossil fuels are burned, no greenhouse gasses are emitted, thereby minimizing environmental hazards. As of date, the initial cost of geothermal heating and cooling is significantly higher than more conventional gas, oil, or heat pump technologies for residential heating and air conditioning needs. As the price of petroleum rises, thereby increasing the operating costs of petroleum-based heating systems, geothermal will become an economically feasible alternative.

One recent development in the utilization of wind energy in the form of wind farms is to move them offshore. Piles are driven into the ocean floor five to ten miles from shore, and conventional windmill turbines are erected on platforms supported by the piles. Underwater cables connect the generators to transformers on shore to convert the voltage to an acceptable level for distribution. While creating some hazards for ships in the area, this application moves the noise and perceived unattractiveness of wind farms away from residents. Regrettably, the environmental impact of such an installation is still being researched.

WIND POWER

Energy from the wind has been used for centuries to propel ships, to grind grain, and to lift water. In farm areas at the turn of this century, electrical power had not been distributed to homes. It was conceived that the wind could be used to provide a mechanical energy not only for water pumps, but also for electrical generators. The major problem was that the wind did not blow all of the time, and then, when it did it blew so hard that it destroyed the windmill. Most of these units used fixed-pitch, propeller blades where the angle cannot be adjusted, propellers as fans. If the angle of the blades could be adjusted, say to a 0° angle, the windmill probably would not be destroyed by the wind. They also used low-voltage, direct-current electrical systems and there was no way of storing power during periods of low-wind speed.

The wind generating plants used today have a system of storage batteries and other components that provide a constant power output even when the wind is not blowing. They also have a 2- or 3-blade propeller system that can be protected, or feathered, during high-wind periods so that the mill will not be destroyed. A simplified wind system is shown in Figure 12-6. Most of these systems are individual units capable of producing 120–volts of alternating current with constant power outputs in the low–kilowatt range. The generator output may be interconnected through a system of series–connected batteries with an automatic, solid-state voltage control that is designed to convert the dc voltage to 120-volt ac. The cost of these individual electrical power-generators is fairly low and these systems provide a complete self-contained, non-polluting, power source. Wind systems would be ideal for remote homes that have a low-power requirement. The have proven to be dependable and have low-maintenance costs. Larger power-generating plants could be possible when located in windy places. These plants would probably be limited in their power output due to the requirement for large storage batteries. Because of the varying wind speeds, it would be difficult to connect two or more units together properly. Compared with other types of power production from alternative energy sources, for example solar power, wind systems are smaller in size per kilowatt output and have a lower initial cost. However, a solar power plant can only produce power during sunshine periods while a wind generator will operate whenever there is wind.

The primary disadvantage of a windmill is obvious. What do you do on days when there is no wind? One answer is the use of storage batteries. Another disadvantage is that windmills generally have a low efficiency, approximately 50%. Although windmill power will probably never be a major contributor to the solving of any power crisis, some individuals have calculated that several large-diameter windmills, with a fairly constant wind, could produce many kilowatts of electrical power. They could play an important role in reducing the use of our natural resources for electrical power production, particularly on an individual basis.

TIDAL POWER SYSTEMS

The rise and fall of waters along a coastal area caused by gravitational forces is the basis for the tidal electrical power production method.

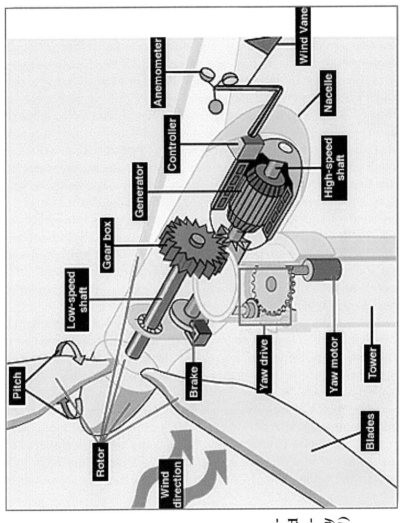

Figure 12-6. The mechanics of a wind turbine. (*Source: National Renewable Energy Laboratory [NREL]*)

There are presently several tidal power systems in operation. Tidal systems would be desirable since they do not pollute the atmosphere, do not consume any natural resources, and do not drastically change the surrounding environment as do some conventional hydroelectric systems.

There are two practical possibilities when constructing a tidal power system to generate electricity. These are the one- and two-basin schemes. A one-basin scheme is where water is allowed to enter the bay of estuary with the turbines and will generate electricity when the tide falls. This scheme has the disadvantage that the power availability will not match the power demand over time. The second method is a the two-basin scheme shown in Figure 12-7. Two basins are separated by a dam containing the turbogenerating units. One basin, called the low pool, is permitted to empty itself with the low tide and is then sealed off to the incoming tides by the use of gates. The other basin, which is called the high pool, is allowed to fill with the incoming tide. The water in the high pool flows through the turbines into the low pool. When the tide begins to ebb, the sea gates of the high pool are then closed. When the tide drops lower than the level of the filling low pool, the low pool's gates are opened to release its water. The objective of this scheme is to increase the versatility of the system by providing for power generation at any time. This advantage over the single-basin technique costs money, because an additional system of dams is required.

The depth of tidal water varies at different times of the year. These depths are determined by changes in the sun and moon in relation to the earth. Tides are predictable since the same patterns are established year after year. A tidal power system would have to be constructed where water in sufficient quantity could be stored with a

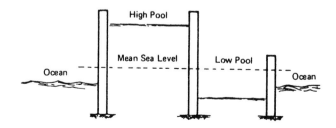

Figure 12-7. Diagram of two-basic scheme.

minimum amount of dam construction. A tidal system could be made to operate during the rise and fall of tides. Also, the pumped-storage method could be used in conjunction with tidal systems to assure power output during peak load times. A potential tidal system site along the United Stated-Canada border has been studied, but the economic feasibility is not promising.

BIOMASS SYSTEMS

Another alternative system being considered as a potential method of producing electrical power is biomass. Biomass sources of energy for possible use as fuel sources for electrical power plants are wood, animal wastes, garbage, food processing wastes, grass, kelp from the ocean, or any plant-derived material. Many countries in the world use biomass sources as primary energy sources. In fact, the United States used some of the biomass sources almost exclusively for many years. The potential amount of energy which could be produced in the United States by biomass sources is substantial also at this time.

Uses of biomass available are ethanol, biodiesel, and biomass power. Both ethanol and biodiesel are substitutes for fossil fuels such as gasoline and diesel fuel. While the fossil fuels have a limited supply, the biomass equivalents, being plant-based, are endless. In addition, biofuels burn cleaner, thereby reducing emissions, greenhouse gasses and pollution. Biomass power is the use of assisting coal-fired electrical generating stations with a biomass fuel. Although the efficiency of a system is somewhat limited, this is touted as a short-term solution to the issues of electricity generation. Research in the areas of biorefineries to reduce dependence on oil is now underway. Plant-based materials such as paper pulp and corn will be used to produce products similar to those produced by the process of oil refining. If proven, this technology could provide fuels in large volume. The hydrolysis of cellulose to make ethanol at a lower cost is also under development. In addition, thermo-chemical conversion methodologies to increase the efficiency of biomass fuels are currently being researched by the US Department of Energy. Questions still exist regarding the use of biomass for the production of electrical power, however the ability of biomass to reduce dependence on fossil fuels such as oil is a reality.

COGENERATION

When electrical power is produced, typically a fuel is consumed to produce steam, and this steam turns a turbine that spins a generator to produce the electricity. Because as little as one third of the original energy spent to produce the heat to generate the steam is transferred to electricity, significant waste occurs in the generation of electricity. However, if the heat energy that was formally wasted is used for other purposes efficiencies can be much higher. Power-producing facilities capable of reclaiming this lost energy are dubbed cogeneration facilities. In fact, anytime heat and power are produced simultaneously, it is titled cogeneration or combined heat and power (CHP). Cogeneration is actually a combination of many technologies designed to trap this steam or wasted heat for other purposes, thereby increasing efficiency to over fifty percent. Typically a cogeneration plant is located near the end-user of the heat and electricity produced, thereby reducing electrical as thermal losses. This heat energy is captured in a heat exchanger placed in typical areas of loss such as in flues or cooling towers, and the reclaimed heat is sent directly to the user. In some instances, materials other than gas, coal, and oil are used to generate heat for steam in cogeneration facilities. Renewable resources such as biomass fuels can be utilized.

MAGNETOHYDRODYNAMICS (MHD) SYSTEMS

MHD stands for magnetohydrodynamics, a process of generating electricity by moving a conductor of small particles suspended in a superheated gas through a magnetic field. The process is illustrated in Figure 12-8. The conductors are made of metals, such as potassium or cesium, and can be recovered and used again. The gas is heated to a temperature much hotter than the temperature to which steam is heated in conventional power plants. This superheated gas is in what is called a plasma state. This means that the electrons of many of the gas atoms have been stripped away to make the gas a good electrical conductor. The combination of gas and metal is forced through an electrode-lined channel which is under the influence of a superconducting magnet that has tremendous strength. The magnet must be of the superconducting type since a regular electromagnet of that strength would require too much power. A superconducting magnet is one of the key parts to this type of generation system.

It becomes a difficult job to keep the conductive channel from being destroyed with high operating temperatures and a high-speed gas flow. Cooling is important and is accomplished by circulating a suitable coolant throughout the jackets built into the channel. Also due to the high temperatures and the metal particles moving at high speed, erosion of the channel also is a critical problem. This particular problem has been eased by using a coal slag injected into the hot gas and metal stream. The coal slag acts to replace the eroded material when it is lost.

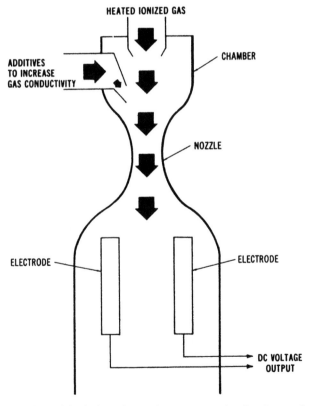

Figure 12-8. Simplified drawing of a magnetohydrodynamic (MHD) generator system.

Pollution is not a big problem. There is a difficulty with the high levels of nitrogen oxides produced. These oxides are the direct result of the high combustion temperatures inherent to the system. Sulfur oxides and ash are also a problem for any plants using coal or oil. An afterburner system has been proposed to eliminate the nitrogen oxides while the sulfur oxides and the ash would be collected, chemically separated and, then recycled.

At this time, MHD generators are primarily experimental. There have not been very many units made and the ones that have been produced have not been in use for a long period. Efficiency is of importance for this type of generation system. While the conventional coal-fired steam generation system is 40% efficient, MHD plants could operate at 60% efficiency. Fuel supplies could be enhanced by using this generation system. At the present, the future of MHD systems for large-scale electrical power generation is questionable.

NUCLEAR POWER

The majority of the time when the word "nuclear" is mentioned, we tend to be uncomfortable. From the various nuclear accidents which tend to make major headlines and the illegal nuclear dumping, leakage, etc., it is understandable. No one can truly say how safe nuclear power is, but the plus-side needs to be addressed.

A positive aspect of nuclear power is the environmental advantage of not releasing large amounts of air pollutants as do fossil fuel plants. Fossil fuel plants also require large quantities of fuel which require many trains, pipelines, or oil tankers. Certain types of reactors use fuel that is in almost limitless supply.

On to the negative aspects, a release of radiation no matter how small still takes place at an atomic power plant. Another major negative aspect is the storage of radioactive wastes. Such wastes are radioactive and extremely hot. These wastes will be stored for tens of thousands of years. Containment vessels tend to deteriorate because of the heat and the radioactivity. Attempts are under way to solidify wastes and store them in underground vaults such as salt mines. Salt is impervious to the flow of liquids.

NUCLEAR FISSION

Nuclear power plants utilize reactors which function due to the nuclear-fission process. Nuclear fission is a complex reaction which results in the division of the nucleus, or center, of an atom into two nuclei. This splitting of the atom is brought about by the bombardment of the nucleus with neutrons, gamma rays, or other charged particles and is referred to as induced fission. When an atom is split, it releases a great amount of heat. A nuclear-fission power plant simulator is shown in Figure 12-9.

In recent years, several nuclear-fission power plants have been put into operation. A nuclear-fission power plant, shown in Figure 12-10, relies upon heat produced during a nuclear reaction process. Nuclear reactors "burn" nuclear material whose atoms are split causing the release of heat. This reaction is referred to as nuclear fission. The heat from the fission process is used to change circulating water into steam. The high-pressure steam rotates a turbine which is connected to an electrical generator. The nuclear-fission system is similar to fossil-fuel systems in that heat is used to produce high-pressure steam to rotate a turbine.

Uranium-235 has proved to be a valuable nuclear fuel, but about 1% of the uranium metal ore mined is of the uranium-235 type. But it has been said that the burning of less than one ounce of uranium-235 is equivalent to the heat produced by one ton of coal.

Nuclear Fuels

A sustained nuclear-fission reaction depends on the proper use of fuel. The most desirable fuels available are uranium-233, uranium-235, and plutonium-239. As we have stated before, uranium-235 occurs naturally. The other two fuels, uranium-233 and plutonium-239, are artificial offshoots of uranium-235.

Breeder Reactors

A reactor that uses uranium-233 and plutonium-239 is called a breeder reactor. During the nuclear reaction which takes place in a breeder reactor, materials that are used in the reaction process are converted to fissionable materials to be used again. In other words, in this type of method there is no waste and the reactor produces its own fuel. The long-range development of nuclear power production may be dependent upon whether or not breeder reactors can be available soon.

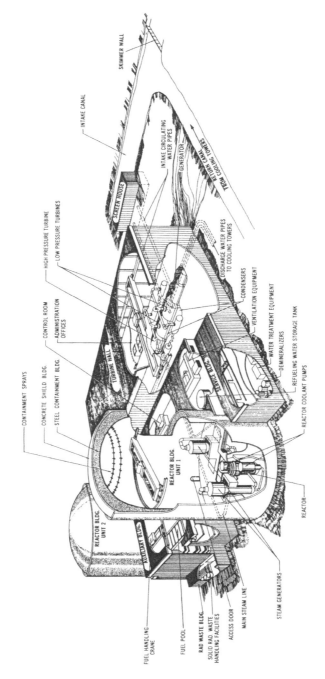

Figure 12-9. An operational nuclear-fission power plant.

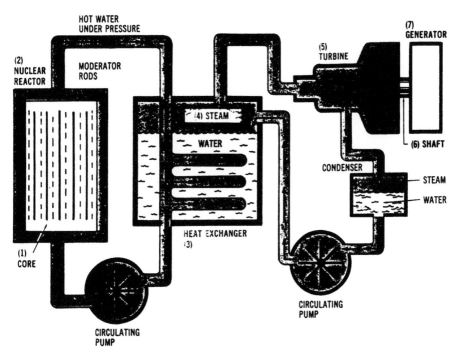

Nuclear fission in the core (1) of the reactor (2) produces energy in the form of heat, which heats water under pressure. The heat from the water in this primary system is transferred to a secondary stream of water in heat exchanger (3) converting it into steam (4), which spins the turbine (5) connected by shaft (6) to generator (7), producing electricity.

Figure 12-10. Drawing illustrating the principles of a nuclear-fission power system.

NUCLEAR FUSION

Another alternative power production method which has been considered is nuclear fusion. Deuterium, the type of fuel used for this process, is abundant. The supply of deuterium is considered to be unlimited since it exists as heavy hydrogen in sea water. The use of such an abundant fuel could solve some of the energy problem related to the depletion of fossil fuels. Another outstanding advantage of this system is that its radioactive waste products would be minimal.

The fusion process results when two atomic nuclei collide under controlled conditions to cause a rearrangement of their inner structure. Following the process, a large amount of energy is released. These nuclear reactions or fusing of atoms must take place under tremendously high temperatures. The energy released through nuclear fusion would be thousands of times greater per unit than the energy from atypical chemical reaction and considerably greater than that of a nuclear-fission reaction.

The fusion reaction involves the fusing together of two light elements being released to form one heavier element and the heat energy being released during the reaction. This reaction could occur when a deuterium ion and another ion are fused together. A temperature in the range of 100,000,000°C is needed for this reaction to produce a velocity for the two ions to fuse together. Sufficient velocity is needed to overcome the forces associated with the ions. The fusion reaction produces a helium atom and a neutron. The neutron, with a high enough energy level, could cause another reaction with nearby ions. A much higher amount of energy would be produced by a nuclear-fusion reaction than a fission reaction.

Future of Nuclear Fusion

Since nuclear fusion is considered to be environmentally safe and would use an abundant fuel, much research is being done to formulate ways of controlling this reaction process. Fusion reactors have not yet been developed beyond the theoretical stages. The problems surrounding the construction of this type of reactor centers around the vast amount of heat produced. The fuel must be heated to a high temperature. The heated fuel must be confined so that the energy released by the process can become greater than the energy that was required to heat the fuel to its reaction temperature. This is necessary to sustain the fusion reaction and to produce continuous energy.

Solutions to these problems and development of economically attractive commercial systems will take time and money. However, it seems very possible that such systems may soon be developed.

Alternative Nuclear Power Plants

A unique concept in electrical power in the 1970s was called a floating nuclear power plant. These plants were proposed to be nuclear-fission plants mounted on huge floating platforms for operations on the water.

The proposed unit would be located on rivers, inlets, or in the ocean. The plant would be manufactured on land and then transported to the area where it would be used. The electrical power produced by the plant could be distributed by underwater cables to the shore. The power lines could then be connected to on-shore, overhead power-transmission lines. The floating nuclear power plants could be mass-produced, unlike conventional nuclear facilities which are individually built.

Floating nuclear plants could have advantages over other power systems. There are obvious ecological benefits. A plant located in the water would have less thermal effect on the water due to heat dissipation over a large body of water. Also, these units would not require the use of land for locating power plants. They would be flexible since they could locate on rivers, inlets, or oceans.

HYDROGEN

Hydrogen is one of the simplest elements and third most abundant element on earth, as found in organic and natural compounds such as water. It is this same element that makes up much of the fuels in use today such as gasoline and combustible gasses in the form of hydrocarbons. Pure hydrogen is high in energy, as proven by the use of liquid hydrogen to propel rockets into orbit since the 1970s. While technologies for the production of pure hydrogen are being developed, they are not likely to produce hydrogen fuel to be burned as in a rocket. At this time, burning hydrogen for the propulsion of a transportation vehicle or generation of heat for electrical power is not considered to be feasible due to the highly unstable nature of the fuel and cost of production. However, a great deal of research is currently underway identifying hydrogen as one of the keys to making fuel cells a feasible source of electricity. Authors and experts in the field have predicted hydrogen will be the basic or central form of energy in the distant future.

FUEL CELLS

Fuel cells are devices that can convert chemical energy to electrical energy, similar to a battery. Unlike a battery that typically has a limited or fixed fuel sources to which derive electricity, fuel cells use an external fuel, such as hydrogen to generate electricity. As long as the fuel is supplied, the fuel cell will continue to generate electricity. This process is pollution-free, quiet, and fairly efficient. An advancement in fuel cell technology is the polymer electrolyte membrane fuel cell. This technology appears to hold promise for electric powered vehicles and perhaps homes. Also under development is a regenerative fuel cell. In a regenerative cell, the reverse of a fuel cell reaction is used, electrolysis. This is significant because the electrolysis of water produces hydrogen and oxygen, two fuels for fuel cells. Once proven, this fuel cell could generate electricity in a closed environment, without the need for additional fuel.

Chapter 13

Energy Cost Reduction / Going Green

INTRODUCTION / GREEN ENERGY CONCEPT

In today's society "green" refers to being environmentally friendly. In terms of energy green refers to conserving energy in order to lessen the impact energy consumption upon the environment. All dwellings, residential, commercial or industrial use energy. This energy is used to maintain our comforts such as heating and cooling, water heating, lighting, and other tasks like running appliances, machines, or other equipment. Because this demand can be 24 hours a day, 7 days a week, any methodologies to conserve energy will result in a direct reduction in energy costs. How much energy costs will rise in the short or long term is impossible to determine. However, one fact is abundantly clear, energy costs are not projected to decline.

A brief study of the graphical representation of almost any energy resource such as natural gas, electricity, and oil over a period of time, all exhibit characteristics of an exponential curve, a curve defined by means of an exponential equation, which can never reach its outermost axis. The end to the rising cost of energy is nowhere in sight. Several forms of energy can be considered green energy. Solar, hydro, geothermal, and tidal forms of energy collection are some of the ways that energy can be gathered with a minimal negative effect on the environment. In addition to gathering energy in a minimally evasive manner, a green building—a structure that maximizes energy use—is an excellent way to reduce human impact on the environment. A green building will garner the most possible benefit from the least amount of energy.

In this chapter, a summary of many methodologies for reducing energy consumption, and thereby reducing funds expended on energy costs are explored. The chapter contains useful tools to aid in the process, as well as actual methods, techniques, and applications that will result in a reduction of energy costs. The chapter is laid out similar to the systems discussed in the textbook where savings in building structure, heating, cooling, lighting, water and other areas of energy consumption are examined in greater detail with a focus on cost reduction and environmental impact.

ASSESSING ENERGY CONSUMPTION / AFFECTING THE ENVIRONMENT

The most likely place to begin a quest for energy savings is with an energy audit. There are two types of audits, a do-it-yourself format and an audit conducted by a professional. While this is the starting point for all most all energy savings analysis, the authors of this book believe that there are additional steps, data gathering steps, which can provide valuable insight as to the energy consumption of the dwelling.

The first tool is an analysis of the energy consumption pattern, or the creation of an 'energy profile'. Commonly referred to as a utility ledger (Figure 13-1), this tool provides an illustration of the consumption from various utilities over time, and results in a total energy cost figure. The final area of the ledger is the energy utilization index. This index is a calculated measure of the efficiency of the building represented on the ledger. This allows direct comparison between buildings or a comparison of energy in the same building over time. Also, many of the costs, quantities, and types of fuel recorded in the ledger are the same data and figures that may be required when performing calculations for energy savings in other energy cost reduction techniques. The second tool recommended is conducting a walk-through building survey (Figure 13-2). By determining the present value of insulation, types of heating and cooling systems as well as lighting demands, comparisons can be made between buildings that have differing systems. Also, as with the utility ledger, data and figures in the building survey may be useful when evaluating other energy cost

Building _____

Gross Area (ft)^2 _____

Year _____

Month	Heating Deg. Days	Cooling Deg. Days	Electricity							Fuel					Total Energy Cost
			KWH	KW Demand		Cost			Check Gas ☐ Coal ☐ Other ☐		Cost			
				Actual	Billed	Total	Per Unit		Quantity	Total	Per Unit			

Figure 13-1. Utility ledger.

Annual Energy Consumption in BTU's:

	Quantity		Conversion Factor		BTU/Yr.
1. Electricity . . .	_____kwh	×	_____3413	=	_____
2. Purchased Steam	_____(M) lbs.	×	_____	=	_____
3. Natural Gas	_____MCF	×	_____1,030,000	=	_____
4. Oil	_____Gallons	×	_____ #2-138,700 #6-149,700	=	_____
5. Other Fuel	_____	×		=	_____
6. Total					_____

Energy Utilization Index

$$EUI = \frac{\text{Total Energy Consumption BTU's/Yr.}}{\text{Gross Area (ft}^2)}$$

$$= \underline{\hspace{3cm}} \text{ BTU's/ft}^2/\text{Yr.}$$

Figure 13-2.

reduction techniques.

After data for each building or site have been obtained, then an energy audit can be conducted. As previously mentioned in Chapter 11, an audit is necessary to point out any potential problem areas in energy consumption of a residential, commercial or industrial user of energy. These audits may be conducted by a professional, or as a do-it-yourself project. A professional energy audit generally consists of a room by room walk-through inspection as well as an analysis of utility bills. If available, professional energy auditors have technologies such as thermographic scans available to inspect areas for heat losses. Surprisingly, professional energy audits are available without charge in many areas. These free energy audits are may be sponsored by the local utility companies or branches of the local, state, or federal government that have missions related to reducing energy costs, particularly for low-income sectors. Also, many excellent energy audit checklists are available from many resources, including the Internet, to aid an individual in performing a self-audit. Typically these audits are performed as a walk-through, with checklists to help in problem identification. Whether professional or do-it-yourself, the purpose of an energy audit is the same, it can identify immediate problems with the current structure, or its systems, and can help prioritize future improvements, with energy conservation in mind.

BUILDING STRUCTURE / SAVING MONEY

Whether planning a new structure or retrofitting, repairing or remodeling an existing structure, there are many areas of improvements to study with energy conservation in mind. While some of the techniques discussed in this section offer greater potential savings than others, it is suggested that a 'whole-house' approach be used when assessing energy needs. While some areas cost little, such as reduced air leakage, and others cost a great deal, such as windows and doors, all entities should be considered when designing or altering a structure for reduced energy consumption. After all, changes in one area of a structure may have impacts or effects on another area, possibly for the good, possibly a negative impact.

General Information

Facility Name _____

Year Built _____

Floors _____

Total Area (square feet) _____

Building Envelope (Indicate condition, construction material and approximate. R-value.)

Walls _____

_____

Windows _____

_____

% of wall space _____

Doors _____

_____

Roof _____

Floor _____

Hot Water

Capacity	Fuel	Temperature Setting

Lighting

Area	Incand. Watts	Fluores. Watts	Other Watts	Total Watts

watts/sq. ft = _____

Heating and Cooling

Type of Unit	Location of Unit	Area Served	Type of Control	Location of Control	Temp. Setting	Regular Maintenance

Figure 13-3. Building survey.

Reducing Air Leaks

The least expensive place to begin in an existing building is the identification and repair of air leaks. Air leaks, or drafts, can account for as much as 30% wasted energy in a residential building. Repair of leaks will reduce the costs to heat and cool the structure, as well as compliment any efforts to insulate. Also, drafty areas are not comfortable for human beings. Below is a list of potential areas for outside or unconditioned air to leak into an area that contains air already treated to a desirable temperature:

- Window frames
- Window glazing
- Door frames
- Door weather-stripping
- Switches and electrical outlets
- Baseboards
- Flooring transitions
- Recessed lighting fixtures
- Ceiling lighting fixtures
- Areas missing drywall or plaster
- Recessed or dropped ceilings
- Attic doors
- Holes for plumbing access (typically under sink)
- Holes for electrical access (through outside wall)
- Fireplace dampers
- Leaks around air ducts

The previous list is far from exhaustive but covers many of the likely areas for air leaks to occur. Many methodologies exist for detecting air leaks, but a simple test would include the closure of all doors, windows and flues, turning off any appliance that relies on combustion, and turning off any fan or apparatus designed for exhaust purposes. After the area is sealed, use a moist hand to feel for leaks. A leak or draft will feel cool to a moistened hand. In lieu of the wetted hand, you may use a candle flame and watch for flickering or the movement of smoke from burning incense. The preceding have a very real fire hazard associated with them and should only be attempted with the utmost concern for fire safety.

Potential leaks can be sealed by caulk, expandable foam or weather-stripping, whichever is appropriate for the application. Great care should be taken when applying expandable foam, because of the ability to expand many times its size at time of dispensing, and requiring a solvent for removal, manufacturer instructions should be followed carefully. In regard to weather-stripping, if the leak is due to a worn weather-strip, select a suitable replacement. Air leaks are not limited to the interior of the structure. Outside the structure inspect any area where building materials meet such as at the foundation, at the ends of siding, all exterior corners and where chimneys attach.

Doors, windows, and skylights are primary locations for energy waste to occur. As previously stated, be sure weather-stripping and seals on doors, windows, and other points of entry are working properly. An old trick to determine if a seal is working properly is to insert a dollar bill between the door or window frame and weather strip. If the currency moves around easily, there is likely a crack large enough to allow leakage. However if the bill resists movement, there is likely an adequate seal. If the structure has an attached garage, the seals around the garage door should also be inspected. Both the seal between the door and the floor and the seal between the door and the frame should be working properly. If no such a seals exists, they should be installed. If the existing doors and windows do not have proper weather-stripping, consider installing an aftermarket sealing mechanism. In certain circumstances, it may be possible that a door is leaking air because of improper closing. In these cases, a spring or hydraulic assist to close the door may be necessary. In some applications, a secondary or storm door may help.

Insulation

Guarding against air leaks and air infiltration is achieved primarily through insulation. As previously mentioned, many wall, ceiling, and floor voids need insulation in order to conserve energy. As a rule of thumb, walls should have R12 and attics R30 in order to insulate properly, however these numbers are simply a guideline. The location of the building and the extremes of outside temperatures, as well as local and state building codes will play a large part in the value of insulation required. However, if the building is 10 or more years old, it likely does not meet current insulating guidelines. Because the cost of insulation is relatively low, retrofitting existing structures (particularly in the attic) is one of the primary methods for energy savings. Insulated suspended ceilings are another methodology, not only do they offer insulation, but they reduce the total volume or space that requires air temperature treatment. To determine if the amount of insulation in an attic is adequate, a quick visual inspection along with a few charts regarding the R-value of various types of attic insulation can provide a good estimate of the R-value or quantity of insulation. Determining the amount of insulation in a wall can be difficult. Inspections behind outlet or switch covers may reveal some insulation, or other methodologies for making small inspection holes can be used, but the only true method to determine if a wall is properly insulated is by thermographic inspection by infrared camera.

Great care should be made when specifying doors and windows for new construction. Not only will these items be major capital expenditures on such a project, but doors and windows can also determine the energy efficiency of the dwelling. While having an expense that is higher at the time of purchase, thermo-pane windows and low-loss doors may quickly pay for themselves in areas where the climate offers temperature extremes. Also, design considerations such as a vestibule can offer energy savings by creating an airlock effect in high traffic areas.

HEATING AND COOLING / SAVING NATURAL RESOURCES

According to the Department of Energy, heating and cooling are responsible for over 50% of the total energy used in a typical home. Therefore, any changes made to increase heating and cooling efficiency will have an excellent payback in energy savings. This unit will examine not only the equipment used for heating and cooling, but the efficiency of supporting equipment as well as factors that influence costs. One significant factor has already been mentioned, the need to reduce air leaks and provide proper insulation. Any energy converted by a heating or cooling system and lost through leakage or improper insulation is effectively money wasted, because the energy is not going toward its intended use. Therefore, it is recommended that along with, or prior to the examination of heating and cooling systems for potential energy savings, that insulation and air leaks be addressed as well. Not only will this let the heating and cooling system be most effective, but should also improve the level of comfort in the dwelling.

Heating and Cooling Equipment

Many factors play a part in the selection, efficiency, and cost of operation of any heating and cooling system. Some systems such as geothermal are more energy efficient by design than other systems. How well insulated and sealed against air leaks also determine how efficient the system will be. The temperature differential between what is desired inside and the temperature outside will play a large part in the energy consumption. Little factors such as thermostat settings play a part. It is estimated that as much as two percent of heating costs can be saved by every degree the thermostat setting is lowered during periods of heating demand, or raised during summer cooling. Programmable thermostats that are capable of setting the temperature to a desired level based on occupancy can also save because the equipment is not operating when the there is no demand.

Surprisingly, the best methodology to improve the efficiency of a heating or cooling system is perhaps one of the simplest and frequently the least expense, regular maintenance. Changing of filters, lubricating equipment, sealing duct and cabinet leaks are all included in maintenance of the system. A dirty air filter in a forced air system could rob as much as 1/5 of the energy from the system. The amount of maintenance needed by systems vary, but manufacturers recommendations should be followed, and it is recommended that a heating and air-conditioning contractor clean and inspect the unit annually. By having a professional clean the system, dirt can be removed from the coils (both inside and outside) that can restrict air flow and significantly decrease thermal transfer. Also, a professional inspection can reveal problems when they are small, and possibly less expensive to repair.

In addition to insulation, proper thermostat setting, and performance of regularly scheduled maintenance, below is a list of other efforts that can reduce heating and cooling costs:

- Clean ducts and registers
- Use ceiling fans to aid in circulation
- Use heat from the sun to aid in winter heating
- Shade the sun to aid summer cooling
- Install tinting or film on windows
- Install automatic valves on radiant heaters
- Install high-efficiency blower motors
- Supplement with passive solar heating
- Install a unit with higher efficiency

Should the building have a heating and cooling system that is more than 15 years old, replacing the system with an energy-efficiency unit might be feasible, depending upon the energy consumption of the existing unit. Consultation with a local contractor should aid in this decision. As a rule of thumb, units with higher efficiency will generally be able to 'pay for' their increased cost of initial purchase within a few years of operation. The size of the system depends on many factors such as the location, square footage of the building, types of doors and windows, and the type of occupancy. Manuals such as Manual J are published to aid professional contractor in sizing systems properly. Also, computer software is available to do the same.

LIGHTING

According to the Department of Energy, lighting is responsible for approximately 10% of the total electrical consumption in a typical home. Because of the recent changes in lighting in the form of compact fluorescent bulbs and LED lighting, it is possible that lighting changes could have a more dramatic effect than ever on total electrical consumption. Perhaps the simplest methodology for saving energy is to turn off supplemental lighting when it is not needed or in use. The rules for turning off lamps varies with the technology, but a good rule of thumb is to turn off incandescent lamps whenever they are not in use and fluorescent lamps if not in use for 15 or more minutes. Other techniques such a installing a lamp of a smaller wattage still apply, but installing the newer technologies may offer greater rewards. However, this unit does not begin with the virtues of compact fluorescent or LED lighting, but a discussion of a light source that is less expensive, daylight.

Whenever possible, incorporate day-lighting to maintain high energy efficiencies. This could be as simple as windows and skylights, light tubes, or a more complex light harvesting and distribution system. There are variables to account for when using daylight, such as heating, cloudy days, and occupancy during night times. Technology exists to sense the amount of daylight in a room, and dim or turn off all or some of the supplemental room lighting. In addition to being a nearly free resource, most people find daylight more pleasing than artificial lighting. To prove this, research the number of varieties of light bulbs and lamps on the market that claim to recreate daylight. Many of the enhancements in the lighting industry at this time deal with the color or hue of the light emitted by the lamp, to mimic the spectrum of colors found in daylight. However, if the area needing lighting is not on an outside wall or the top floor, technologies generating artificial lighting are likely the most economical alternative, and will be discussed next.

With the introduction of new lighting technologies, the Department of Energy has determined that approximately 50% or more savings can be realized in terms of reduced consumption. Moving from incandescent lamps to fluorescent and compact fluorescent lamps is a part of this. Also included are solar powered lamps for exterior lighting and the application of LEDs. In general, it is a good practice to examine each installation of incandescent lighting to determine if it is a good candidate for replacement by fluorescent or HID lighting. If changing the source of the light does not detract from efficiency, visual quality, or amount of light needed for the designed activity, it is likely that the application can derive energy savings from a change-over. In addition, the design of the lighting, such as offering the proper amount of light, better reflection, as well as efficient controls such as dimmers and occupancy (motion) sensors will decrease the electrical demand. As with heating and cooling, light systems should be properly maintained to maintain top efficiency. Removal of dust and dirt, cleaning reflectors, and replacement of yellowed lenses or other factors that inhibit the light from reaching the intended area should also be considered.

Other factors that influence the lighting in a particular area must be considered. Examine the wall, floor and ceiling surfaces. Are they light in color or reflective? If not, possibly changing the texture or color of a surface to reflect more light could offer a savings in electrical consumption necessary for lighting. Be aware that taking this suggestion to the extremes can cause an undesirable problem called glare. This can be a particularly annoying problem if it occurs in a work area, but may be distracting in a hallway or seldom used space. One way to spread light around is to add diffusers on the fixture or area where daylight is entering the building, such as a skylight. Also, it is important that such light diffusers be kept clean.

Another simple way to save energy in a space in which only one area is typically used to carry out a task, such as a desk in an office, is to use task lighting. Task lighting is a general name given to any lighting that is specifically aimed at an area for a task, rather than spreading the light through a room where it may not be needed. Task lighting can be as simple as a desk lamp, or special lighting fixtures installed permanently over the task area. The savings is simple, by not lighting unneeded areas, energy is saved, while lighting is optimized where needed.

Another important area in which to assess lighting is the possible elimination of unnecessary fixtures, or relocation of these fixtures to a more appropriate area. Examine the area to determine if the lighting is appropriate. Is there a fixture located where light is not necessary, such as near windows, or areas that have little to no tasks that are bathed in light. If so, it may be possible to eliminate the fixtures. One way to determine this is to remove the light temporarily and reassess the effect the change has on the area. Incandescent can be unscrewed or disconnected, but fluorescent fixtures require more than a bulb removal. To assess changes here, the ballast must be disconnected by a qualified electrical expert, or temporary shading installed in the fixture. Removal of lamps in a fluorescent fixture is not permitted, because the ballast is still 'on' and using energy even with the removal of the bulbs. Other assessments of lighting in areas might reveal an ability to relocate light fixture from areas in which they are not needed to other areas where increased lighting is desired. Another method to obtain more light from existing fixtures is to determine if retrofitting the fixture with a reflector is a valid option. Not only will this provide more lighting, but save on the cost of replacing the entire fixture. One last technique is to simply replace the fixture with one that promotes energy savings.

There are many techniques to determine the type, intensity, color, style, and number of fixtures necessary to provide lighting in various areas of residential, commercial, retail, and industrial spaces. It is beyond the scope of this book to cover all of the possible techniques, however it is suggested that in a new construction or remodeling situation that such a strategy be employed. Whether driven by a handbook or a suggested layout from a lighting retailer or supplier, be certain that not only enough light is present for a job to be performed effectively in the area being lighted, but that energy savings are also considered. Make certain that factors such as daylight and LED lighting are at least assessed for viability.

WATER / SAVING OUR VALUABLE RESOURCE

Demands for potable water are increasing, and the resources at this time are limited. As demand increases, the costs of having fresh water continue to escalate, any technique or technology to increase water efficiency will become increasingly desirable. The easiest method to conserve water, and save on costs associated with the water, is to eliminate waste. It is estimated that up to 15% of the water used by a household is wasted, either by leakage or simply not being used properly, and sent down the drain.

Leakage

According to the California Urban Water Conservation Council, the number one area in which water is wasted in leakage. While a dripping faucet is likely the most common leak, there are many others that can be considered. For instance, any water device you notice is running slower, or taking longer to complete a task may be suffering from a leak. Even small leaks have large wastes associated with them. It is estimated that a leak that only drips one time per second can waste over 3000 gallons annually. This unit will look at some of the primary areas of leakage, and when possible, offer hints on how to repair or correct the problem.

The most likely leak location is a faucet in a sink, tub, or laundry area. While faucets come in many styles, they all have a common feature, the need to seal water until it is called for by turning the faucet on. Some faucets utilize washers, others seals, and others cartridges, but all perform the same basic function. In general, faucet repair parts are inexpensive. Often a handy, do-it-yourself person can be complete the repair successfully. Should it be necessary to have a plumbing professional repair the faucet, it is advisable to have the faucet rebuilt or assessed for replacement because the service fee will likely be the highest line item on the bill in the event of a faucet repair.

Toilets are a silent area in which a water leak can occur. While it is possible to leak from the fill valve, most toilet leaks occur around the valve that separates the tank from the bowl. Leaks such as these may be nearly impossible to detect, but there are signs. The most obvious sign is a toilet that 'runs' or refills the tank without use. Over time, a water leak from the tank to the bowl will cause the tank to empty, and at some point the water fill valve will need to refill the tank. When a toilet is not being used, it should not need to refill the tank. Other hints include the sound of water running in the toilet when all the valves should be closed or movement of water in the bowl of the toilet when there should be none. Different methodologies to enhance the detection of a leak such as putting food coloring in the tank of the toilet and watching for a color change in the bowl have met with limited success. As with faucets, repair parts for toilets are relatively inexpensive, and a qualified do-it-yourselfer can often perform the repair successfully. Also, if the toilet is not compliant with the 1.6 gallon per flush (GPO rating, it may be time to replace the toilet.

When water is mounted below the level of the ground, such as a pool, spa, or irrigation system, leaks can be very difficult to detect. However, inspecting for loose fittings and seals in such devices as well as perpetual wet spots or unexplained dampness in nearby ground can lead to the detection of leaks. Because underground leaks are so difficult to detect, they are some of the largest. For example, a leaky pool liner might drop the level of water in a pool less than 1/8 of an inch per day, but the effective leak would be in excess of 350,000 gallons per year. Professional leak detection for underground installations using equipment such as submersible cameras, supersensitive microphones, air, or dyes is available, but this professional service comes with a price.

Generally speaking, if a leak is suspected, turn off all devices in the dwelling that consume water, including those that fill automatically. Make a note of the reading on the water meter and leave the area for some time. Make sure no other persons enter the area, or if they do, make them aware that a leak detection process is underway, and make certain they use no water. After a period of time, re-check the water meter to determine if any water has been consumed. If the meter has moved, it is likely that the water consumed is through a leak. One final note, typical water meters are not extremely sensitive to water consumption and may measure in hundred gallon increments, so waiting a sufficient amount of time for the leak test is critical.

Having an abundant supply of clean water is a concern that will continue to grow in the 21st century, and it is likely that new technologies to reduce water consumption, reuse existing water, and technologies to recycle or reclaim dirty water will emerge in this century.

Certainly computer control of water dispensing will help, but the next gain will likely be in the area of reuse of water. Just as many municipalities require car wash owners to recycle a large portion of the water used in the facility, techniques to use gray for flushing, irrigation, and other tasks that do not require the water to be clean in order for the task to be accomplished will hit the mainstream in due time.

Water Heating

Using proper techniques to heat water is another area in which energy savings can be realized. Water heaters are the typical areas of interest, however other devices such as waterbeds, hot tubs, and dishwashers also are responsible for heating water. In order to save energy, it is recommended that the thermostat for water heaters be set at 130° to 140° Fahrenheit. Blankets that increase the insulating value of the water heater are also recommended. In addition, not all water heaters are created equally. When specifying a new or replacement water heater, pay particular attention to the energy rating. As with most devices that are large consumers of energy, the small increase in price of an energy efficient unit will more than pay for itself over the life of the unit. Also, other techniques to save on energy for water heating are suggested. If a waterbed is being used, make certain the blankets are placed back on the bed when it is not being used. Similarly, spas and hot tubs should be covered when not used, and settings on these devices may be turned down if it is known they will not be in use for a period of time.

Water Conservation

One guaranteed effective method to save on the cost of water is simply to use less of it. Below is a list of some commonly accepted methods for water conservation.

- Install aerators
- Install low-flow showerheads
- Take shorter showers
- User energy saver settings on appliances
- Wash clothes in cold water
- Purchase an efficient clothes washer
- Only run dishwashers when at capacity

- Water plants only when needed
- Use graywater to water foliage

Graywater is the water that goes down the drain from sinks, showers, tubs and washing machines. This water can be reused for irrigation and landscape watering. While this list is far from exhaustive, it does make the point that energy spent on delivery, treatment, or heating of water is truly wasted, and great care should be exercised to keep this type of waste to a minimum.

ELECTRICAL APPLIANCES / IMPROVED DESIGN

Any appliance or device that converts energy to do a job or perform a task is a potential point of energy waste. Because many of the devices discussed here use electricity as their primary energy source, it is important to understand how energy consumption for an electrical device is calculated, thereby aiding in the discussion of energy savings.

Calculating Electrical Consumption

The standard unit of electrical consumption is the kilowatt-hour or kWh. This is determined by multiplying the power consumption of a device in watts by the number of hours used, and dividing this product by 1000. The formula looks like this:

(Consumption in Watts x Hours in Use)/ 1000 = Kilowatt-hour (kWh)

If the number or hours in use is calculated per day, this could be multiplied by the number of days in use and the annual kWh consumption could be found. Once multiplied by the local rate for electricity, the annual cost of operation could be calculated. For example:

A fully loaded desktop computer with a 17 inch CRT monitor may consume as much as 400 watts of power. The computer is used for studying and games 4 hours each night, 7 days a week. If electricity costs 8¢ per kilowatt-hour, what is the energy cost of operating the computer?

(400 Watts x 4 hours/day)/1000 = 1.6 kWh of consumption per day
1.6 kWh x 365 days = 584 kWh of annual consumption
584 kWh x $.08 = $46.72 of annual electrical energy costs.

As shown in the example, a common device can actually have a significant impact on the total cost of energy

household. If a desktop computer is responsible for approximately $50 of annual costs, imagine the costs of major appliances such as water heaters, stoves, or electric dryers. The key to understanding the cost of operating appliances is in determining the wattage or power consumption of the appliance. The Department of Energy offers these suggested ranges of power consumption for the following appliances (in Watts):

- Aquarium = 50-1210
- Clock radio = 10
- Coffee maker = 900-1200
- Clothes washer = 350-500
- Clothes dryer = 1800-5000
- Dishwasher = 1200-2400 (using the drying feature greatly increases energy consumption)
- Dehumidifier = 785
- Electric blanket- Single/Double = 60/100
- Fans
  — Ceiling = 65-175
  — Window = 55-250
  — Furnace = 750
  — Whole house = 240-750
- Hair dryer = 1200-1875
- Heater (portable) = 750-1500
- Clothes iron = 1000-1800
- Microwave oven = 750-1100
- Personal computer
  — CPU -awake/ asleep = 120/30 or less
  — Monitor -awake/ asleep = 150/30 or less - Laptop= 50
- Radio (stereo) = 70-400
- Refrigerator (frost-free, 16 cubic feet) = 725

- Televisions (color)
  - 19" = 65-110
  - 27" = 113
  - 36" = 133
  - 53"-61" Projection = 170 - Flat screen = 120
- Toaster = 800-140
- Toaster oven = 1225
- VCR/DVD = 17-21/20-25
- Vacuum cleaner = 1000-1440
- Water heater (40 gallon) = 4500-5500
- Water pump (deep well) = 250-1100
- Water bed (with heater, no cover) = 120-380

The exact power consumption of any appliance should be found on the manufacturers tag identifying the electrical characteristics of the device. Figure 13-4 is a tag from a small appliance. Notice it is rated for use at 120 volts, 60 Hertz, and consumes 40 watts of power. Other devices such as window air conditioners, treadmills, air handlers for heating and air conditioning, and other large consumers of energy are not on the list, but should be considered when calculating the annual energy costs of a dwelling. In fact, any appliance that uses electricity for heating or cooling, a pump, a fan, or other devices that contain electric motors should be considered as a possible target for

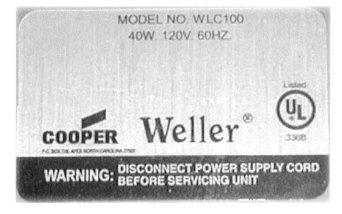

Figure 13-4. A typical product tag identifying electrical characteristics.

increasing efficiency by lowering its power consumption.

Just as mentioned in the unit on heating and cooling, it is important to maintain all electrical equipment in order to keep energy consumption low. Refrigerators, freezers, dryers, etc. need to be inspected for leaks around doors and seals. Worn belts, hoses, and other consumable materials need to be replaced as necessary or as recommended by scheduled maintenance suggested by the manufacturer. Portable heaters and window air conditioners should be clean and free of dirt and dust that hinder thermal conductivity. Also, if an appliance that is a large consumer of energy fails, consider replacing it with an energy efficient unit.

Just as it is true with other energy consumption devices, the savings over time from purchasing a more efficient appliance can likely pay for the device over the life of the device. Be sure to compare not only price but energy consumption when purchasing major appliances such as refrigerators, freezers, heaters, ovens and dryers. The easiest place to find such information is on the EnergyGuide Label. Established by the Federal Trade Commission, this bright yellow and black label represents an estimate of the products energy consumption. By law all new appliances are required to have such a label, and can be a useful tool in comparison shopping. However, calculations on actual energy consumption may vary from the label, therefore it is considered only as a guideline. Figure 13-5 is a mock-up of an EnergyGuide label for a refrigerator. Note the comparison of the model represented to similar models in energy consumption in kWh, and the estimated annual operating expense.

LANDSCAPING / GREEN SPACE

There are many, many methodologies to employ in which energy can be saved. Regrettably it is beyond the scope of this text to provide an exhaustive list of all technologies and techniques, but one simple and easy technique that often goes unnoticed remains, landscaping. When designing or redesigning the landscaping for a residential area, consider plants that comply with the following:
• Appropriate for the climate
• Produce shade or windbreak

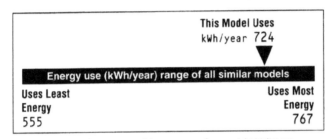

Figure 13-5. An EnergyGuide Label

- Conserve water

    Plants that are appropriate for the climate will likely not need as much water to sustain life, but be more beautiful year-round. Plants that provide shade will reduce the need for cooling in the summer months, thereby saving energy, and likewise, plants that provide a natural windbreak will reduce the need for heating in the winter months. Also, any plant that is capable of conserving water will reduce the need for constant watering or irrigation, as well as be easier to maintain.

ENERGY MANAGEMENT PROGRAM / FOR GOING GREEN

This chapter has presented many ideas where energy savings can be achieved. It is now the responsibility of the owner or manager of each installation, area, home or facility to implement some sort of energy management program. Such a program with an organized system of priorities can aid in decision-making by focusing efforts on areas offering energy savings. There are many methods by which an energy management program can be developed or implemented, the approach taken in this chapter is somewhat generic in an effort to let each user customize the model presented here to suit the needs of the varying areas of need.

While many energy management programs exist, many follow this simple seven-step process listed below. In some instances the steps may be combined or blurred, other applications may require the steps be performed in a slightly different order. The primary steps are:

1. Identification of the need
2. Assessment of the needs
3. Develop an action plan
4. Implement an energy team
5. Execute the plan
6. Evaluate the plan
7. Spread the word

Identification of the Need

Just as with many projects, the identification of the needs is the place to begin an energy management program. Either by using suggestions from this chapter, or a list provided by a third-party vendor or the government, begin identifying areas in which energy savings could be realized. Leave nothing out, heating and cooling, building structures, lighting, water, landscaping, look at everything. Identify all areas that could improve energy efficiencies and log each. At this point, do not assess each change for economical efficiency, which is step number two.

Assessment of the Needs

After all the needs have been identified, assess each for potential energy savings, costs of implementation, and other variables associated with correcting the need identified in step number one. Approach this as a feasibility study. Generate estimates of costs to install the proposed change, and estimate the cost savings. Prioritize each need based on the estimated payback. Recall that some projects will have impacts on other projects, so be sure the assessment includes these secondary effects, positive or negative. Likely those projects or needs with a greater payback should be placed higher in priority on an action plan, if monies (existing or requiring financing) to implement the change are available. It is also possible that other needs identified will cost more to correct than the estimated cost of the loss, and it may be decided to remove these items from the action plan.

Develop an Action Plan

Once a prioritized list of needs have been established, an effective plan of action will insure the list is implemented properly. In addition to identification of the needs, the action plan should include timelines or goals for implementation, persons or departments responsible for the changes, and other project management decisions. This process step will vary depending upon the size of the organization, and the number of needs identified and assessed. Should the action plan represent a large project, there are several software tools to aid in scheduling tasks.

Implement an Energy Team

The concept of an energy team is to allow easier facilitation of the changes in the action plan to be implemented. If the project is small in scale, this step may be eliminated, or if the project is larger, this step may need to be moved up in the process. It may require a team to identify the needs in a large facility. Should an energy team be needed, a director or manager with the power and authority to implement the action plan should lead such a team. Be certain that other stakeholders are represented on the team. Major changes might affect building engineering, traffic flow, or result in temporary losses of some services. If the project is of a large scale, vendors, contractors, and perhaps the utilities themselves should hold a seat on the team.

By including all stakeholders on the team, better and more effective communication can result.

Execute the Plan

This step is straightforward, though an individual or team leader, implement the action plan. This step of the process can take on various timelines, depending upon the length of the projects identified in the plan. It is the responsibility of the leader to insure the projects are being completed on schedule and on budget. Any variations should be noted, and those tasks not yet implemented that can be affected by such variations should be updated.

Evaluate the Plan

After a task has been implemented, it should be evaluated to determine if the projected savings was indeed an accurate reflection of the actual savings. This can be a an ongoing or formative assessment after each project, or a more summative assessment addressing large portions of the action plan, or the entire action plan. As before, as discrepancies are identified, these changes should be fed back into the tasks remaining on the action plan, and each should be re-assessed for possible reprioritization. Other possible areas of evaluation may include how the plan is being executed, what has gone well, what could be improved.

Spread the Word

Should a particular project reap great rewards, share the information with others. Speak to owners of similar facilities, and let them know about the project, and the realized savings. The responsibility for saving energy not only benefits the end-user, but large scale implementation of such projects will have a profound influence on our planet and the quality of life.

Appendix 1—
Basic Electrical Symbols

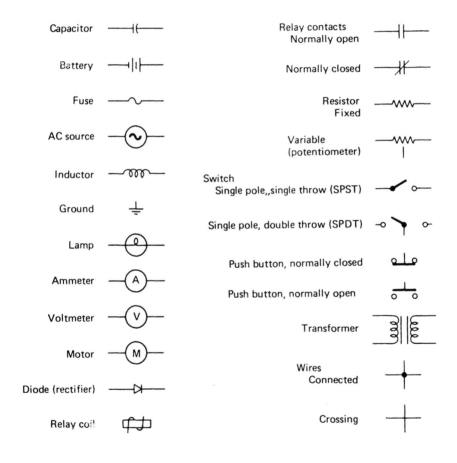

Capacitor

Battery

Fuse

AC source

Inductor

Ground

Lamp

Ammeter

Voltmeter

Motor

Diode (rectifier)

Relay coil

Relay contacts
Normally open

Normally closed

Resistor
Fixed

Variable
(potentiometer)

Switch
Single pole,,single throw (SPST)

Single pole, double throw (SPDT)

Push button, normally closed

Push button, normally open

Transformer

Wires
Connected

Crossing

Glossary

Absolute pressure –A pressure value measured from a perfect vacuum.

Absorber – A device of a mechanical refrigeration unit that takes in a vapor solution applied to it from the evaporator after it boils out of the generator.

Absorption cooling system – A cooling cycle in which refrigerant is moved from an area of strong solution by heat, then condensed in a condensing unit, and expanded by pressure in an absorber. The refrigerant evaporates to provide cooling and is absorbed by the remaining solution. The solution is then pumped back into the condenser.

Active solar system – A system that has solar collectors, a thermal solar storage system, and a transfer medium which converts sunlight into thermal energy.

Adsorption cooling system – A cooling cycle in which air to be conditioned is dried, causing an increase in temperature. The air is then cooled in its dry state to reduce its enthalpy (total heat content) and passed through a humidifier which cools the air at its reduced heat content.

AFUE – Annual Fuel Utilization Efficiency. A measure of furnace efficiency obtained by comparing the heat delivered to energy supplied to the furnace.

Air infiltration – The leakage of air into or out of a building (dependent on outside temperature compared to inside temperature) through cracks and around doors and windows.

Alternating current (AC) – The type of electrical power that is produced at electrical power plants and distributed to homes and industrial and commercial buildings. Single-phase ac is used for homes and other lower-power applications. Three-phase ac is used for higher-power industrial and commercial applications.

American Wire Gauge (AWG) – A device used as the standard measuring system for electrical wire in the United States.

Ampacity – The electrical-current-carrying capacity of a conductor used for electrical wiring.

Ampere (A) – The fundamental unit of current

Annual energy index (AEI) – The ratio of the total energy use per year (in millions of Btu) in a building area (in thousands of square feet). The annual energy index is a method of showing the energy use of a building compared to other buildings. AEI is expressed in thousands of Btu per square foot per year.

Alternate energy sources – Nontraditional methods of extracting energy which are now under experimentation. This includes coal gasification and liquefaction, oil shale extraction, nuclear fusion, Magnetohydrodynamics (MHD), wind geothermal, biomass, solar, and ocean power.

Ambient temperature – The environmental temperature in which a solar energy system operates.

American Society of Heating, Refrigeration, and Air-Conditioning Engineers (ASHAE) – This group develops standards which deal primarily with the energy conservation of buildings.

Apparent power (VA) – The amount of power that is delivered to a circuit or system. The product of applied voltage times the current (volt-amperes).

Atom – The smallest unit of which all matter is composed.

Atmosphere – The gaseous envelope or portion of the earth.

Atmospheric pressure – A gas pressure that is always present and is exerted on the earth by the weight of the atmosphere. Atmospheric pressure at sea level is 14.7 psi.

Auxiliary system – Equipment that uses some form of energy other than solar to supplement the output of a solar system. This provides the necessary backup during periods of low solar output.

Balance point – The place where electric heat is supplied to the system.

Ballast – A coil of wire used to develop a high-voltage discharge for starting some fluorescent and high-intensity-discharge (HID) lamps.

Barometer – An instrument that measures atmospheric pressure.

Biomass – An alternate energy fuel source. Examples of this are wood, methane, gasohol. Any fuel derived from a biological source.

Bourdon element – The element is used to indicate temperature values on a chart or calibrated scale. A coil or flat tubing with its end sealed. Increases in pressure tend to cause the element to straighten.

Branch circuit – The conductors of an electrical distribution system which extend beyond the final overload protective equipment (fuses or circuit breakers).

Branch drains – Smaller sewer lines that connect alternate fixtures to the main soil stack. These lines are smaller, depending upon the number of fixtures being served.

British thermal unit (Btu) – The amount of heat needed to raise the temperature of 1 pound of water 1 degree Fahrenheit.

Building energy performance standards (BEPS) – A set of standards developed by the U.S. Department of Energy which specify the minimum energy conservation performance required of a building and the systems used in a building, including the building envelope, heating, cooling, ventilating, and lighting systems.

Building envelope – The external surfaces of a building structure which are exposed to the climate. These parts of the building including the roof, walls, doors, and floors.

Building Officials Code Administration (BOCA) – A group that develops codes that are used as a guide for energy conservation in buildings. The codes are recommended for adoption by state or local governments for building code administration and enforcement.

Bus duct – A metal enclosure which contains large high-current copper or aluminum electrical conductors.

Calorie – The amount of heat needed to raise the temperature of 1 gram of water 1 degree Celsius.

Candela – See Candlepower.

Candlepower – Luminous intensity expressed in candles.

Celsius scale (C) – A thermometer-scale which water has a freezing point of 0° and a boiling point of 100°.

Chiller – A mechanical refrigeration unit that contain a compressor, condenser, control valve, and evaporator coil.

Circuit breaker – A device used to open an electrical circuit automatically when excess current flows through the circuit in which the breaker is connected.

Coagulant – A chemical that forms a small, fluffy mass called a floc in natural water. Floc particles are like snowflakes that float around in water during the purification process. These particles are used in removing suspended matter in water and carry it to the bottom of the sedimentation tank.

Coefficient of utilization – The ratio of lumens of light on a work area to the total light produced by the lighting system.

Cogeneration – Production of energy of more than one kind at the same time, such as electrical energy produced together with heat energy in the form of steam or hot water.

Cohesion – When individual molecules cling together.

Collector – A device used to absorb solar radiation and transfer thermal energy to a heat-transfer medium. This normally takes the form of a flat-placed design and mounted on a roof.

Color rendition – The effect of color appearance of objects with reference to their appearance while subjected to a reference light source.

Combustion – The process of changing a fossil-fuel into heat.

Component Performance Standards (CPS) – Design standards that specify the minimum energy performance required of a building envelope, HVAC system, lighting, and other building systems.

Compound gauges – Instruments that measure pressures both above and below atmospheric pressure.

Compressor – A unit in a mechanical refrigeration system that contains a pump used to remove refrigerant vapor from the evaporator coil.

Condensation – The resulting effect when a vapor or gas is changed into a liquid.

Condenser – The segment of a mechanical refrigeration unit that helps passage of cool air. Air forced to pass around the outside of a coil in the condenser removes heat from a gaseous refrigerant that circulates inside of the coil. Lowering the temperature of the gas causes it to change into a cool liquid that is circulated through the system.

Conduction – The transmission of heat or of electrical energy by the motion of particles of the conductor. For example, if one end of a solid metal bar is placed in a flame, it will not take long for the other end of the bar to become hot.

Conductor – A material usually formed into round or rectangular shapes that will allow electrical current to flow through it.

Conduit – A round, pipelike metal or plastic enclosure for electrical conduction.

Conservation of energy, the law of – Energy can neither be created or destroyed. (This is also the first law of thermodynamics.)

Contrast – The difference in brightness between an object and its background.

Control – The part of a system that manually or automatically regulates the system.

Convection – The process of transmitting heat through a fluid such as a liquid or a gas. For example, water boiling on a stove top.

Cooling load – A measure of the amount of energy needed to cool a building. It can be determined on a monthly, yearly, or seasonal basis. It is calculated by multiplying the overall U-value (thermal conductance) of a building by the total surface area of the building times 24 hours per day times the number of cooling degree-days of the time period being calculated.

Corporation cock – a cutoff valve used to control the flow of water from the corporation main to the building.

Corporate main– The main artery that distributes water from carrying infectious diseases. This part of the system is owned by the investors or a municipal agency.

Critical solar radiation–The minimum solar radiation needed to maintain the absorber plate of a collector at the temperature of the heat-transfer fluid that enters it.

Current – The movement or flow of charged particles, called electrons, through a conductor.

Degree–day (cooling) – The degree-day value for any particular day is the difference between the average daily temperature and a temperature of 65°F. For example, for an average daily temperature of 82°F, the number of cooling degree-days is 17 (82°F – 65°F).

Degree-day (heating) – The degree-day value for any particular day is the difference between 65°F and the average daily temperature. For example, for an average daily temperature of 45°F, the number of heating degree-days is 20 (65°F–45°F). Heating degree-days provide an indication of how severe the weather has been for the winter season.

Demand factor (DF) – The ratio of average electrical power used to the maximum of electrical power used over a specific period of time.

DF = average per (kW)/peak power (kW)

Demand meter – A meter used to monitor the demand of a system. This compares to the peak power of the system to the average power.

Design temperature difference (DTD) – The difference between the outside temperature that is expected to occur several times per year and the desired indoor temperature.

Dewpoint – A temperature at which condensation occurs.

Diffuseness – A measure of the direction of light to a surface. Diffuse light comes from several directions.

Diffuser – An object which is placed in front of a light source to control the amount of light emitted.

Diffuse reflection – Reflection of light in all directions.

Diffuse solar radiation – Solar radiation that occurs when sunlight has been scattered by particles in the atmosphere such as moisture in clouds so that it arrives at the earth's surface from several directions.

Direct-beam solar radiation – Sunlight that follows a direct path from the sun to a point on the earth's surface.

Direct current (dc) – The type of electrical power that is produced by batteries and power supplies and used primarily for portable and other specialized applications.

Disinfection – A water purification operation designed to keep water from carrying infectious diseases. It can be accomplished by heating or adding chemicals to water. Chlorine is commonly used to disinfect drinking water.

Domestic hot-water system – A system that produces potable water.

Dry-bulb temperature – The actual temperature of the air. The dry-bulb temperature is measured with a glass thermometer and is not affected by the moisture in the air.

EER – Energy Efficiency Rating. Obtained by dividing the Btu rating of a unit by the power it consumes (in watts).

Efficacy – The ratio of usable light produced o the total energy input to a system or fixture. It is expressed in lumens per watt produced by a lamp.

Efficiency – The ratio of the illumination of an area to the electrical energy used to light the area.

Electrical circuit – An arrangement of electrical devices which has a voltage source, a closed wiring path (conductors), and a load that converts electrical energy into another form of energy.

Electron – A subatomic particle that contains a negative charge.

Emergency Building Temperature Restriction Program (EBTR) – A federal program designed to achieve energy conservation through reduction in room and building temperature.

Energy – The ability to do work. For example, electrical, mechanical, light, and heat energy.

Energy audit – A method by which a person or persons go through a building and identify energy and/or cost savings that would result if energy conservation changes were made in the operation or if modifications could also be called an energy assessment.

Energy conservation – The efficient use of systems that consume energy.

Energy efficiency ratio (EER) – An index of the efficiency of a building system. The sticker that accompanies the majority of appliances. The higher the rating on the sticker the more energy efficient the equipment.

Energy management – A continuous planning process which is used to accomplish the efficient use of energy in a building or system.

Enthalpy – A term used to express the total heat content of the air in units of Btu per pound.

Evaporate – When a liquid is heated and changes into a gaseous state.

Evaporator – The part of an air conditioning unit that assists warm air passing through a coil structure and causing liquid refrigerant to boil and change into vapor. This is similar to an automobile radiator.

Fahrenheit scale (F) – A thermometer scale having fixed points of 32° (freezing point of water) and 212° (boiling point at normal pressure).

Fault – An electrical short circuit.

Feeder – The conductors of an electrical distribution system between the electrical service entrance of a building and any overcurrent protective equipment.

Filtration – A water purification operation designed to remove bacteria and suspended particles after coagulation and sedimentation. Sand and gravel are commonly used in most water filtration operations.

Fixture vents – Interconnecting lines between specific fixtures and the main soil stack for maintaining proper trap pressure.

Floodlighting – A lighting system designed to light a large area. Ordinarily, luminaries that can be aimed in any direction are used for floodlighting.

Flow control device – A device that is designed to automatically regulate liquid refrigerant flow between the condenser and the evaporator coil.

Footcandle – The amount of illumination a distance of 1 foot from a standard candle light source. One footcandle is equal to 1 lumen per square foot.

Footlambert – The brightness of a surface that reflects 1 lumen per square foot.

Force – A push or a pull. The work done per unit of distance.

Fossil fuels – Fuels, such as coal, oil, natural gas, and propane, which are derived from fossil deposits in the earth.

Frequency – The number of cycles (revolutions) per second of an AC generator, measured in hertz (Hz).

Fuse – An electrical overcurrent device that opens a circuit by melting when an excess current occurs.

Gas – Matter that has no definite volume or shape and little cohesion between molecules taking its form from its container. Typical examples of gas are oxygen, hydrogen, and neon.

Gasification – The conversion of low-grade coal into gaseous fuel that has a higher heat content and burns more cleanly.

Gas pressure – This type of pressure occurs when gas molecules are heated and become excited

Geothermal power – An alternative power source using the molten mass of liquid and gaseous matter within the earth's crust. This high powered steam turns large turbines to generate electricity.

Generator – A system of machine used to convert mechanical energy into electrical energy.

Graywater – Water drained from some sinks, showers, tubs and washing machines for non-potable uses such as water for plants.

Ground – There are two types of grounds: system grounds and equipment grounds. System grounds are current carrying conductors used for electrical power distribution. Safety grounds are not intended to carry electrical current but to protect individuals from electrical shock hazards.

Ground fault – An accidental connection to a grounded conductor.

Ground fault interrupter (GFI) – A device used in electrical wiring for hazardous locations to detect fault conditions and respond to open a circuit rapidly before shock occurs to an individual or equipment.

Head pressure – A pressure that is developed as a result of a fluid being elevated to a position above other components. for example, below its surface, a body has head pressure.

Heat – A form of energy that is considered to be a measure or an indication of quantity. It is generally measured in British thermal units (Btu).

Heat exchanger – A unit used to transfer thermal energy from one medium to another.

Heat gain – The amount of heat a building gains due to several sources, such as air infiltration, people inside of the building, lights, and sunlight. This heat is usually removed from occupied buildings by the cooling or ventilation system.

Heating load – A measure of the amount of energy needed to heat a building during a monthly, yearly, or seasonal period. It is found by multiplying the overall U-value of a building times the total building surface area times 24 hours per day by the number of heating degree-days per time period being calculated.

Heat loss – The amount of heat a building loses due to several sources, such as around doors, windows, through walls, floors, and ceilings.

Heat pump – A heat/air-conditioning unit which has the capability of reversing the flow of refrigerant so that the unit's output may be used for either heating or cooling.

Heat–transfer medium – A fluid that is used to move thermal energy from one point to another (usually water or air is used).

Heat of vaporization – The heat that is released when vapor condenses.

HFC – HydroFluoroCarbons, a compound commonly found in refrigerants.

Horizontal branch lines – Horizontally installed pipes that carry waste material from fixture branches to the main soil stack. Each line must be pitched downstream to produce an unobstructed flow.

Horsepower – The basic unit of mechanical power. A single horsepower is equivalent to 550 foot-pounds per second or 746 watts (about 3/4 of a kW).

Hot conductor – A wire that is electrically energized or "live" and is not grounded.

HTML – HyperText Markup Language. A format for publishing on the World Wide Web.

Humidity – A measure of the amount of water vapor in the air.

Hygrometer – A moisture sensor that measures the moisture content of the air through some physical change.

HVAC System – A common abbreviation for the heating, ventilating, and air-conditioning system. This total mechanical system provides comfort heating, ventilation, and air conditioning for a building.

Incident solar angle – The angle between the sun's rays and the normal (perpendicular to the surface) to an area on the earth's surface.

Indirect lighting – A lighting system in which luminaries distribute 90 to 100% of the light emitted in an upward direction.

Impedance – The total opposition to current flow offered by an ac circuit that is measured in ohms.

Insulating glass – Two or more panes of glass separated by air space.

Insulator – A material that will not conduct electrical current under normal conditions.

Kinetic energy – The energy that a body has because of it motion.

Kinetic theory – A theory of matter stating that anything moving does a certain amount of work and possesses some energy.

kWh – Kilowatt hour. The measure of one kilowatt (1000 watts) of energy over one hour time.

Langely (Ly) – A measuring unit for the amount of solar energy intensity.

Latent heat – Heat that is added to or removed from a substance that brings about a state change but not a change in temperature.

Latent heat of condensation – Heat that is absorbed by a gas or vapor and released once it has cooled.

Latent heat of vaporization – When liquid undergoes a change from a liquid to a vapor and no temperature change is involved.

LED – Light Emitting Diode. A type of semiconductor device that emits light.

Life-cycle cost – The cost over the life span of a piece of equipment including operation and maintenance costs.

Liquefaction – The process of changing a solid into a liquid.

Liquid – Matter that has a definite volume but does not have a specific shape. It conforms to the shape of the container that it is placed. Common examples of liquids are water, oil, alcohol, and gasoline.

Liquefaction – The conversion of low-grade coal into liquid fuel which has a higher heat content and burns more cleanly.

Liquid receiver – Part of a refrigeration system that is basically a storage tank for a pressurized refrigerant.

Load – Part of a system that converts one form of energy into another form of energy. An example of a load is an electric motor. The motor converts electrical energy into mechanical energy.

Lumen – The amount of light falling on a unit surface, all points of which are a unit distance from a uniform light source of 1 candela. Essentially, it expresses the amount of light output from a source.

Luminaire – A fixture designed to hold lamps and produce a specific lighting effect on the area to be lighted.

Lux – Lumens per square meter.

MCM – Unit of measure for large circular conductors that is equal to 1000 circular mils.

Magma – The molten mass of liquid and gaseous matter within the earth's crust.

Main cleanout – An access connection o the main drain located near the building sewer line. This access connection permits drain cleaning between the building and the corporate sewer line.

Magnetohydrodynamics (MHD) – An alternate energy power source that relies on the flow of conductive gas through a magnetic field. The flow through the field causes a direct-current voltage to be generated.

Maintenance change – A modification in the operation or construction of a building system which involves little or no cost for the purpose of energy conservation.

Maintenance factor – The ratio of maintained footcandles to initial footcandles produced by that fixture.

Main soil stack – A primary flow feedline for waste material from each fixture group. This line generally extends vertically from the main sewer line to the roof stack vent.

Manometer – A pressure-measuring instrument.

Mass – The quantity of matter that a body contains.

Matter – Any material that makes up the world. A material that occupies space and has weight. Examples of this would be air, water, wood, metal, stone, paper and living things.

Mechanical refrigeration – A process that uses components in an interconnected system to transfer heat from one object or space to another.

Melting point – The temperature at which a solid turns into a liquid.

MERV – Minimum Efficiency Reporting Value. A rating system for air filtration.

Molecule – A particle of which all matter is composed.

Motor – A machine that converts electrical energy into rotary mechanical energy.

NEMA – National Electrical Manufacturer's Association, an organization that establishes standards for electrical equipment.

Neutral – A grounded conductor of an electrical circuit which normally carries current and has white insulation.

Neutron – A subatomic particle which is electrically neutral and has no charge.

Nuclear-fission – Power is conducted through the tremendous heat created by colliding atoms. When water is applied to the heat, the water turns to steam and the steam in turn moves a giant turbine to generate electricity. The fuel used uranium.

Nuclear-fusion – An alternative power production method that is related to nuclear fission but uses heavy hydrogen as a fuel. This fuel is not mined but exists in sea water. Radioactive wastes that are prevalent in the fission process are reduced

Nuclear power – An alternative energy source that uses the power of an atomic collision to produce heat and to generate electricity.

Ohm – The basic unit of resistance and impedance in electrical circuits.

Ohmmeter – A meter used to make resistance measurements.

Overcurrent device – A device such as a fuse or circuit breaker which is used to open a circuit when excess current flows in the circuit.

Overload – A condition that results when more current flows in a circuit than it was designed to handle.

PSI – Abbreviation for pressure per square inch

PSIG – Abbreviation for pressure per square inch gauge. This term refers to the psi reading of a particular gauge.

Panel – A box that contains a group of overcurrent devices which supply branch circuits.

Parallel circuit – An electrical circuit that has all components connected across the power source and in which the voltage is the same across each component.

Passive solar system – An approach to solar energy that depends on the architectural design of a building. This system takes advantage of proper orientation, placement of windows, insulation, and the building itself is used to absorb and collect heat. No collectors, motors, fans, pumps, or other mechanical devices are used to accomplish thermal energy transfer.

Path – The medium through which energy travels from a source to a load. A type of path is the electrical wiring used to distribute power in a building.

Payback – The time required for the cost of energy saved as a result of a purchase or modification to equal the total cost of that purchase or modification.

Payback period – The time required to recover the initial cost of an energy-saving purchase or modification.

Photovoltaic cell (Photocell) – A cell usually made of silicon that converts light energy into electrical energy.

Pilot light – A small burning gas or oil flame that is used to light the gas burner or jet of an oil-burning heater. The pilot flame generally heats a device called the thermocouple. The current produced by the thermocouple hold an electric valve open in the main gas-supply line of the burner assembly. If the pilot goes out, the thermocouple cools, and this causes the main gas valve to remain closed until it is relighted

Plumbing – The practice, materials, and fixtures used in the installation, maintenance, and alteration of equipment connected to sanitary sewers, venting networks, and freshwater supply systems.

Plumbing fixture – A receptacle or device that is connected permanently to a water distribution system. It generally utilizes supply water in its operation and discharges wastewater into the sewage system.

Potable water – Water free of impurities in amounts that will cause disease or harmful physiological effects. This water must conform to bacteriological and chemical quality standards outlined by the U.S. Public Health Service Drinking Water Standard.

Power – The rate at which work is done or energy is converted.

Power factor (PF) – The ratio or true factor (watts) converted in a circuit or system to the apparent power (volt-amperes) delivered to the circuit or system.

Pressure – A force that is exerted on a specific area.

Proton – A subatomic particle which possesses a positive electrical charge.

Pyranometer – A meter used to measure the amount of solar radiation, both direct and diffused, which is received on the earth's surface.

Quad – One quadrillion (10^{15}) Btu's.

R-value (thermal resistance) – A measure of an insulating material's resistance to the movement of heat through it. The R-value is the inverse of the U-value (R=1/U). The higher the R-value is, the less heat will be transferred through the building.

Raceway – A channel used for enclosing electrical conductors of a power distribution system, such as a conduit or metal wireway.

Radiation – A process by which heat is transferred through the motion of waves. For example, heat radiates from the sun in waves.

Radiation pyrometer – An instrument that responds to the amount of thermal energy radiated from the surface of an object.

Radiometer – Any meter used to measure solar radiation.

Rankine cooling system – A cooling cycle in which heat is added to a liquid causing it to evaporate. The vapor is then expanded in an engine to produce power for rotating a compressor. The vapor from the engine is condensed and then pumped back where the cycle is repeated.

Reactive power (var) – The quantity of unused power which is developed by reactive components (inductive or capacitive) in an ac circuit or system. This unused power is delivered back to the power source, since it is not converted to another form of energy. The unit of measurement is the volt/ampere (var).

Reflectance – The ratio of the light reflected from an object to the light falling onto that object. This term also means the ability of a material to reflect solar radiation.

Reflected solar radiation – The sunlight reflected by the ground and objects on the ground.

Reflection factor – The percentage of light reflected from a surface and is expressed as a decimal.

Refrigerants – Chemical compounds that are used as the heat-transfer media in mechanical refrigeration systems. They are alternately compressed and condensed into a liquid or vapor during the refrigeration cycle. They should have a low boiling point, mix well with oil, and be nontoxic.

Refrigeration – A process by which unwanted heat is removed from a selected space or object and kept at a temperature that is lower than its surroundings. This can be achieved with ice, snow, chilled water, or mechanical refrigeration.

Relative humidity – A measure of the moisture in the air at a certain temperature compared with the amount that it could contain if it were saturated. A relative humidity value of 100% indicates that the air is fully saturated with moisture. The abbreviation for this term is RH.

Resistance – The opposition to flow. The opposition may refer to electrical current, fluids, or mechanical motion.

Retrofit – Changes made in a building to make it more energy efficient.

Reverse-flow air-conditioning principle – If a unit air conditioner were turned around in a window during its operational cycle, it would be extracting heat from the outside air and pumping it indoors. This condition is the operational basis of the heat pump.

Risers – Vertical pipes or lines that distribute water between floors or from one level to a higher level.

Romex – A trade name for nonmetallic sheathed electrical cable.

Saturation – This occurs when air holds as much moisture as possible at a specific temperature and pressure.

Sedimentation – A water purification operation wherein suspended particles settle to the bottom of a tank or storage area, where they can be removed.

Secondary stack – An alternate vertical line that carries waste water from remote fixtures other than water closets. The upper part of the stack extends through the roof and responds as a secondary stack vent.

SEER – Seasonal Energy Efficiency Ratio. By including costs for the entire season, it is an improvement over the EER rating. Obtained by dividing the Btu rating of a unit by the power it consumes (in watt-hours).

Semiconductor – A material that has a value of electrical resistance between that of a conductor and an insulator, and is used to manufacture solid-state devices such as diodes and transistors.

Sensible heat – Heat added to or removed from a substance that causes a change in temperature but not a change in state.

Series circuit – An electrical circuit has components connected end to end to form a single path which has the same current through each component.

Service entrance – A set of conductors used to deliver electrical power from a transformer to a building.

Sewage – Any liquid waste matter in suspension or solution.

Short circuit – A fault condition that results when direct contact (zero resistance) is made between two conductors of an electrical system, usually by accident.

Sling psychrometer – An instrument that measures both the dry and wet bulb temperatures of the air to determine relative humidity.

Solar-assisted heat pump – The use of a heat pump in conjunction with a solar collection system to provide heating or cooling of a building.

Solar azimuth – The angle formed by the sun to the horizon and due South.

Solar building – A building that utilizes either an active or passive solar system for heating or cooling.

Solar constant – A reference value which is the total energy emitted by the sun's rays at a specified distance and unit of time.

Solar degradation – Deterioration of materials or equipment caused by exposure to sunlight.

Solar domestic hot-water supply – A system that uses solar collection to produce thermal energy to produce potable hot-water.

Solar energy – An alternative energy source harnessing the sun's rays for heating, cooling, or electrical generation.

Solar insolation – The solar radiation upon a surface which consists of direct beam radiation, and reflected radiation.

Solar spectrum – The various wavelengths of electromagnetic radiation in which the sun's energy exists. This includes ultraviolet, visible, and infrared light.

Solar storage system – The part of a solar energy system used to store thermal energy so that it can be used when needed.

Solid – Matter that has a definite volume and physical shape. Examples are stone, glass, metal, wood, and paper.

Solidification – The process of changing matter from a liquid to a solid.

Source – Part of a system that supplies energy to other parts of a system. Examples of a source are: (1) an emergency generator that supplies electrical energy to a building in case of power failure, or (2) a battery that supplies energy to start an automobile.

Specific heat – The quantity of heat needed to raise 1 gram of water by 1 degree.

Specific humidity – This term refers to the actual moisture contained in a given amount of air.

Stack vent – An extension of the main soil stack through the roof. Air is admitted to the system through this line and gases are expelled.

State change – When matter is brought about by altering the energy level of individual molecules.

Superheating – Additional heat produced by compression of hot refrigerant vapor.

Switch – A device used to open or close an electrical circuit.

Temperature – Level of the intensity of heat. It is measured on a definite scale in Fahrenheit or Celsius.

Therm – A measure of gas fuel equal to 100,000 Btu.

Thermal energy – Heat contained in a material which is caused by the movement of molecules.

Thermistor – A solid-state device used to detect changes in temperature.

Thermocouple – A device that converts heat into an electric current.

Thermometer – A sealed glass or plastic tube with a bored hole and filled with mercury or alcohol used to measure the temperature of a particular area or directly to an operating system with the use of a probe.

Thermostat – The control device of a water heater or building that responds to temperature changes by altering the primary source of heat.

Tons of refrigeration – A unit indicating the capacity of a refrigerator unit to produce the equivalent cooling effect of melting 1 ton of ice in 24 hours. Capacity values are 200 Btu per minute, 12,000 Btu per hour, or 288,000 Btu per day.

Torque – Mechanical energy in the form of rotary motion. For example, a motor has torque.

Totalizing flowmeter – An instrument used in water systems that responds to the total flow that moves past a given point during a specific period. The water meter of a building is a typical totalizing flowmeter.

Transmittance – The ability of a transparent or translucent material to pass solar radiation. This can be done directly, through glass, or be diffused.

Trap – A fitting or device that provides a liquid seal to prevent the emission of sewer gases without altering the flow of liquid sewage.

True power (watts) – The amount of power that is converted to another form of energy in a circuit or system. It is measured with a wattmeter.

U-value (thermal conductance) – A measure of the amount of heat an insulated area of a building or an insulating material will conduct. The total U-value of a building structure is its overall thermal conductance in Btu heat loss per square foot of building area per hour per degree Fahrenheit ($Btu/ft^2/hr/°F$).

Vacuum – A space containing no matter. A space in which all of the air or other gases has been removed.

Valve – Control devices designed to stop, start, check, or throttle the flow of water through pipes and fixtures. Float valves, cock valves, cutoff valves, and flow control valves are commonly used in water systems.

Venturi – A burner mixing tube of reduced diameter located in a natural gas burner of a furnace.

Volt – The basic unit of voltage or potential difference in an electrical circuit.

Voltage – The potential difference or electrical pressure across two points in a circuit.

Voltage drop – The reduction in voltage caused by the resistance of conductors of an electrical distribution system, which causes less voltage in the end of an electrical circuit.

Water service line – A buried line of pipe that connects the main, the water meter, and the building interior.

Watt – The basic unit of electrical power. The amount of power converted when 1 ampere of current flows under a pressure of 1 volt.

Watt-hours (Wh) – A measure of power consumed in a building for a given number given number of hours.

Watthour meter – A meter that monitors electrical energy used over a specific period.

Wattmeter – A meter used to measure the electrical energy that is converted in a circuit or a system.

Weather stripping – plastic, metal, or felt strip used to seal spaces between windows or doors and their frames. This reduces air infiltration through the spaces.

Weight – The weight of a body is directly proportional to its mass and inversely proportional to the square of its distance from the center of the earth.

Wet-Bulb temperature (WB) – The moisture content of the air.

Wind Power – An alternative energy source that uses wind turbines to provide power to generators.

Work – Work is accomplished when an applied force moves a body or mass through a measurable distance.

Work plane – A level at which work is usually performed. A horizontal plane 30 inches above the floor is used for lighting design unless otherwise specified.

XML – Extensible Markup Language. A format for delivery of electronic documents, usually via the Internet.

Zone – An area of a building which the HVAC system supplies through one control device. For example, a thermostat.

Index